BEFORE THE
BEGINNING

BRIEFINGS *is a new series of short paperback books by experts in their field on complex subjects of major importance, in a style and at a length that will make them intelligible to the non-specialist reader.*

The books will, in other words, provide authoritative Briefings *on the kind of subject which readers will regularly see referred to in the press, or hear mentioned on radio or television without being fully aware of their meaning. These are nevertheless subjects about which some basic knowledge is vital for an understanding of the contemporary world.*

Themes and topics covered will include Feminism, Education, Cosmology, Medical Ethics, Political Ideology, Structuralism, Quantum Physics and Comparative Religion among others.

Authors for each subject will be leading specialists or academics from major universities, laboratories, research institutes or industries, worldwide.

BEFORE THE BEGINNING

Cosmology explained

George F.R. Ellis

BRIEFINGS
Series Editor: Peter Collins

BOYARS/BOWERDEAN
LONDON. NEW YORK

First published in 1993 jointly by the Bowerdean Publishing Co. Ltd. of 55 Bowerdean St., London, SW6 3TN and Marion Boyars Publishers of 24 Lacy Rd., London SW15 1NL and 237 East 39th St., New York, NY 10016

Distributed in Australia by Peribo Pty Ltd., Terrey Hills, NSW 2084

The **BRIEFINGS** series is the property of the Bowerdean Publishing Co. Ltd.

A CIP catalogue record for this book is available from the British Library.
A CIP catalog record for this book is available from the Library of Congress.

ISBN 0 7145 2970 2 Original Paperback

Designed and typeset by Screen Text DTP, Guildford
Printed and bound in Great Britain by Itchen Printers Ltd., Southampton.

CONTENTS

Preface

PART I: ISSUES AND FOUNDATIONS

PART II: THE SCIENTIFIC UNDERSTANDING

PREFACE

This book is about the relation of humanity to the Universe. The original title was "The Universe: Cradle of our existence". That title indicates quite accurately much of what I want to convey. However I was then persuaded to go for a more snappy title, which still conveys much of its theme. That theme will undoubtedly be controversial; in writing the book, I have tried to steer a careful course between major traps and abysses on either side of the chosen path. I hope that even where you judge that I have failed to avoid these errors, you will at least find the journey stimulating.

Because of the need to keep the length of the book within strict limits, what follows cannot give a systematic presentation of all the underlying issues, neither can it present an adequate historical survey of previous approaches. The aim is to give an overview of a possible approach, a picture of the whole but painted with broad brushstrokes. In effect, it is an indication of a research programme that would aim to fill in the details and answer many of the questions that must inevitably be left unanswered here. To make up for this brevity, I give at the end of the book a good number of references where more detailed discussions of the various aspects may be found. These sources are referred to in the main text by a number enclosed in square brackets, for example [1].

Clearly what is presented here builds on the thoughts of many others, both personal friends and authors whose works I have read. You are too numerous to mention individually; I am indebted to you all. However one specific event particularly helped development of the approach presented in this volume: I thank members of the Vatican Observatory/CTNS (Centre for Theology and the Natural Sciences) Conference held at Castel Gandolfo in September 1991 [6] for their papers and discussions. I am particularly grateful to Bill Stoeger and Nancey Murphy for illuminating comments that helped shape this book. I thank Francis Wilson, John de Gruchy, Rosemary Elliott, Peter Spargo, Helen Zille, and Neil Berens for detailed comments on the manuscript.

George Ellis
Department of Applied Mathematics, University of Cape Town;
also Interdisciplinary Laboratory, SISSA, Trieste, Italy, and
Mathematics Department, Queen Mary and Westfield College, London.
11 August, 1992

All Truth is a shadow except the last, except the utmost;
yet every Truth is true in its own kind.
It is substance in is own place,
Though it be but shadow in another place
(for it is but a reflection from an intenser substance;)
and the shadow is a true shadow,
as the substance is a true substance.

Isaac Penington (1653)

PART ONE

ISSUES AND FOUNDATIONS

CHAPTER ONE

THE BIG QUESTIONS

This chapter introduces the major themes considered in this book: a broad view of the nature of the Universe and its relation to humanity.

Cosmology is the science that studies the physical structure of the Universe. Building on the results of other sciences (particularly physics and astrophysics), it has led to major new understandings about the nature and evolution of the matter around us; and this in turn has led to an understanding of the physical conditions that make life possible on earth. However the aims of the present science of cosmology are much narrower than implied in the anthropological use of the term, where Cosmology refers to an overall world-view that throws light not only on the structure and mechanisms of the Universe, but also on its meaning. We can characterise some of the big questions that have concerned humanity since the dawn of consciousness, as follows (cf. [1]):

> * The nature of the physical Universe: What is it
>   made of and how does it work.
> * The question of creation: How do things originate?
> * The issue of the final state: What will happen in
>   the end?
> * The place of humanity in the Universe: How does
>   all this relate to us?
> * The meaning of existence: What is the purpose of
>   it all?

These absolutely fundamental issues are all the concern of the broader understanding of Cosmology. The modern scientific theory has made great strides on the first issue, and presented many interesting and important theories concerning the second and third. However it has, - for perfectly good reasons - very little to say on the last two; certainly nothing remotely like enough to satisfy those who want more than a completely reductionist understanding, i.e[1] an approach that aims to reduce all explanation to nothing more than the action of the laws of physics (whose nature and existence is taken for granted). As will be explored in this book, such a view cannot by its nature take seriously the breadth and depth of human experience. On the positive side, it is precisely by limiting its scope and adopting a reductionist view within its domain of application that modern science has had such enormous success within that domain. The downside is that it cannot therefore by itself provide a satisfactory overall worldview, even though this is sometimes attempted. The failure of such attempts, and the realisation that major areas of human concern lie outside the scope of the modern scientific vision, go some way to explain a present widespread turning away from science towards various a-scientific or even anti-scientific worldviews (see e.g. [2]).

The theme of this book (in common with various others recently published, for example [3-6]) is that the time has come for the scientific approach to be extended to a more profound synthesis that attempts to return to the broader issues, considering the nature of cosmology and the relation of humankind to the Universe in a way that attempts to throw some light on the big questions without losing the essence of what has been attained through the scientific approach. This thesis is put forward with some trepidation, realising that many will believe the basis for such a step is lacking and the enterprise foolhardy in view of (a) all the evidence for the social and cultural basis of our belief systems; (b) the wide divergence of opinions on philosophical issues; and (c) the failure of so many previous attempts. The main text will attempt to meet these arguments, the view being that knowledge has advanced sufficiently in recent times, particularly in terms of defining the strengths and limits of science and the scientific method, so that it is possible and even appropriate to consider the issue anew.

In order to attain its aims, as well as taking account of the discoveries of modern science, such an attempt must be prepared to consider philosophical and practical issues that are of importance in everyday life. In particular, it cannot avoid taking a serious look at issues both of an ethical and religious nature (without prejudging the outcome of that examination); for these play an important role in many people's lives, and make fundamental truth claims we cannot simply ignore. I realise that there are many who will immediately react against this assertion. However a blank refusal to consider whether significant discoveries could be made by taking these areas seriously is unscientific, for it prejudges the issue without considering the evidence for and against it. The unsatisfactory nature of many religions and philosophies does not mean that the whole area is without significance, but rather that, as in the case of science, one must learn to distinguish between the fruitful and unfruitful ways of approaching understanding in this area. Indeed here the scientific approach can throw some light, helping to attain a broader unified basis for understanding that can be applied in most areas of life, providing it is applied in an appropriate way in each area.

I realise that some of my fellow cosmologists will feel that I am way overstepping the mark of the subject in raising some of the issues discussed here. To resolve this, it may serve a useful purpose to distinguish two different intellectual disciplines. Suppose we use cosmology (lower case 'c') to refer to the technical subject of physical cosmology and all that this entails (Chapter 5 of this book, and the foundational material for that study contained in Chapters 2-4); and that we use the word Cosmology (Capital 'C') to refer to the broader study that undertakes to look at the issues referred to in Chapter 8. Then Chapters 6 and 7 certainly will be included in Cosmology; different proponents of theories of cosmology may or may not wish to include this material in the scope of their deliberations. Given this convention, many possible objections will fall away when it is clearly stated that this is a book on Cosmology (including sections on cosmology, but as components of the larger study).

PART I of the book ('Issues and Foundations') defines the major questions which are its concern, and outlines the approach that will be adopted in tackling them. It is concluded by Chapter 2, which considers foundational issues in attempting the kind of approach envisaged.

PART II (Chapters 3 to 6) presents the present scientific understanding of the Universe. The scientific method has led to a tremendous growth in knowledge of the way matter is organised and functions. The chapters in this part give a brief outline of some of the most important of these discoveries, Chapter 3 focusing on the underlying fundamental laws, Chapter 4 on the functioning of complex systems, and Chapter 5 on the environment within which all this takes place (the Universe around us). However this understanding is uncertain in some aspects, and limited in others; Chapter 6 discusses some of these uncertainties and limitations.

PART III (Chapters 7 to 9) turns to the main theme of humanity and the Universe, providing a broader viewpoint than the purely scientific one, but built on the foundations laid by the previous chapters. Chapter 7 introduces the Anthropic question, i.e. the puzzling fact that the physical Universe seems fine-tuned to permit the existence of complex physical organisms, including human life. A purely scientific approach to understanding this feature is found to be inconclusive. Chapter 8 introduces a broader view of the problem, that attempts to take seriously important aspects of the human condition such as the foundations of ethics. Chapter 9 concludes the book with a comparison of the two major ways of understanding the Anthropic issue. To some extent this suggests answers to the 'big questions' posed in Chapter 1.

I have had to decide whether to use some technical words in this presentation, or to try to avoid them altogether. I have chosen the first course, because while it is a nuisance to the reader having to learn their meaning if it is unfamiliar, it is essential to have some label for each important concept; one may as well use the usual one, for this opens up access to the literature of the subject. If you come across a technical word that is unfamiliar to you, I suggest looking for it in the Index; there is a fair chance that I have given a brief definition of the idea at some point in the text.

CHAPTER TWO

UNDERSTANDING THE UNIVERSE

This chapter considers the basis of the scientific approach to understanding the physical universe, and how this can be broadened into an approach to understanding in general. This provides a good background for the scientific sections in Part II, and lays the foundation for the more broad-based synthesis proposed in Part III.

The scientific and the religious viewpoints are but two of the various ways humanity has tried in the quest for understanding and the search for a pattern of meaning [1,7-10]. We cannot attain understanding without such a search because of one of the fundamental aspects of the situation that confronts us: namely, the *hidden nature of reality*[1] .

Many of the scientific aspects of the world are far from obvious; we cannot without considerable effort and perspicacity deduce the nature of the chemical elements, the fundamental forces that bind matter together, and so on. We can easily appreciate the majesty of mountains and the beauty of flowers, but only with skilled experimentation determine that they are made of a combination of carbon, nitrogen, oxygen, and other elements. Furthermore when we do succeed in understanding physical aspects of reality, its nature may be quite unexpected (for example, the essence of relativity theory and the nature of quantum mechanics).

It would hardly be surprising then if a similar feature were to apply to other aspects of life. If the mechanisms of physics and chemistry that make life possible are hidden, it is possible that any fundamental underlying moral order and purpose that there may be, is hidden too. We can easily see if someone has a fine appearance or beautiful clothes, but much less easily determine what moral character and ethical nature underlies this surface appearance. Nevertheless it may be that discerning what is the good life and the right way to live is also possible, but only with effort and on the basis of accumulated wisdom, and the answer obtained may not be the obvious one.

Thus one of the prime issues is determining the essence and scope of the hidden nature of reality. It is commonly agreed that it incorporates those features that underlie scientific laws. The question is how much wider its scope may be. Furthermore, as this hidden nature of reality is one of the fundamental conditions under which we live our lives, it is therefore one of the features we should eventually try to explain when forming a coherent view of the Universe.

2.1 Different views

The initial attempts to relate to this hidden aspect of nature may be broadly characterised as *magical* or *superstitious* approaches: it is assumed either that if one wishes hard enough for something it will happen, or that if one adopts various ritual practices intended to bring about an effect, it will occur, whereas in fact they have no causal relation to it and cannot influence it. The evident inefficacy of this approach, coupled with obvious injustices that have often resulted from its practice (witch-hunts leading to the death of innocent people, for example) have then led to the development of two diverging world views.

On the one hand the *religious* view has evolved serious approaches to moral issues and to possible understandings of ultimate reality; and on the other the *scientific* view has with tremendous success tackled the issue of immediate causes, leading to an extraordinarily effective understanding of how things work in terms of physical cause and effect. Indeed one of the major triumphs of science is precisely the understanding that because the sequence of natural events is governed by regular laws of behaviour[2], each physical event has causes that can be determined by appropriate investigation; and desired final effects can be achieved by organising an appropriate set of initial conditions. This establishes the firm relation between physical cause and effect.

The achievements of science are undeniable: it has led to discovery of the physical basis of nature and laid the foundations for the huge explosion of technology enabling us to order our lives and control our environment in ways previously unimaginable. In the face of this enormous success, other world views, and in particular the theological one that held sway for many hundreds of year, have been forced to retreat: they have had to modify their truth claims, abandoning to science much previously claimed territory. This has led to a large move away from religion on the part of thinking men and women, also reacting against many of the horrors and injustices perpetrated in the name of religion over the centuries.

However there has recently also been a reaction against science in many quarters, on the one hand because of its perceived contribution to environmental degradation and the development of weapons of mass destruction, and on the other because many see it as based on an outlook that dehumanises and denies the value of the individual [2]. In many cases this has led to what is essentially a return to magic, in the form of astrology, arcane effects ascribed to pyramids or crystals, and so on. These are bound to disappoint eventually, as they are not based on real cause and effect relations. However the fact that people turn to them is evidence that many are seeking a world-view with elements of humanity and hope not provided by the present dominant scientific one. Sometimes the effort is dressed in pseudo-scientific terms, wanting to inherit some of the mantle of scientific success but avoiding the kind of logic and indeed hard work demanded to make good that hope.

For those not following this magical route but who want to take seriously the insights both of science and of some form of broader philosophical or religious world view, there are two choices: to go for separate worlds, a divided mind that uses completely different, indeed probably contradictory, approaches and criteria for different parts of life (for example, the scientist who uses quite different kinds of logic and evaluation of evidence in daily life than in his or her professional work); or to try for a unity of view in some scheme that transcends the viewpoint of either. The latter, broader view is what will be explored here.

In adopting this stance, we are adopting a specific set of values that underlie the approach. Indeed however we may hide it, there is no such thing as a value-free approach to these issues. Broadly speaking, the values adopted here include an attempt to find a description that reflects the truth insofar as we are able to achieve that aim, in particular therefore taking seriously the discoveries and viewpoint of modern science; but they also acknowledge the worth and significance of the broad range of human activity (including aesthetic and moral choices, personal life decisions and experiences) and therefore insist that these must be taken as seriously as the empirical data of scientific experiments. Only in this way can we achieve a truly humanitarian world view.

The danger of the approach is that we will end up with a theory that is completely relative, finding its roots strongly tied to our present perspective and therefore bound to be regarded as wrong by those with different cultural views, or found inadequate in years to come when our perspective has changed. We have to take this risk: being aware of the danger, we take what safeguards we can against falling into the trap of constructing a local or ephemeral theory, purposefully looking for those features we believe are fundamental rather than superficial or transitory. Having made the decisions to do the best we can we plunge in, accepting that there will inevitably be some cultural and temporal bias in our views. This need not undermine the worth of the search for understanding, or the value of what we find; indeed in many cases we can argue strongly we have determined in a culture-free way some invariant aspect of underlying reality [3].

2.2 The common base

In each area of understanding, whether we make the fact explicit or not, our understanding is based on *mental models of reality* that will, to a greater or lesser degree, accurately reflect the nature of some aspect of reality. They may be simple ideas, perhaps comprised in a single label (the word 'cat', for example, conjures up a whole model of appearance and behaviour), or complex theories (e.g. the theory of relativity, or Jungian psychology). Each such model will have a *range of applicability*, specifying both the *set of phenomena* that are its concern, and a (usually restricted) *set of conditions within that domain* that it is supposed to explain. For example Newtonian mechanics explains the motion of physical bodies, provided they do not move at speeds close to the speed of light; a theory of psychology may describe the ordinary behaviour of people at work, but not that of psychopaths.

Theories can never be absolute: indeed the understanding they give will always be partial, because no model can circumscribe within itself the full nature of reality. Thus theories are always subject to revision. Hence the key element underlying the approach proposed is that, whatever the field of application, growth in understanding is based on a *creative proposal of theories to explain reality*, but always founded on the combination of *scepticism*, i.e. the willingness to doubt the current orthodoxy, and testing, i.e. checking that orthodoxy, or any proposed alternatives, against reality in the whole variety of ways that is practical. Indeed this is the foundation of all learning, through the *basic learning cycle* [11]. In essence, we

1. set up a hypothesis on the basis of present knowledge;

2. work out its consequences (in the process checking that it is coherent, making logical sense), and then

3. test its consequences against evidence (if possible performing experiments that can check if it is true or not).

Then we return to step 1, reconsidering the hypothesis and modifying the theory on the basis of the new evidence available (and any new ideas that may have come up).

Of course in reality life is more complicated than this. The hypotheses are set up within a pre-existing framework of understanding that will be based on a cultural and temporal viewpoint; and a theory is made up of a complete set of interlocking hypotheses and assumptions, which are tested as a whole. Thus it will in general not be obvious what it is that needs altering to make a better theory than the present one. Nevertheless the power of observational tests underlies the examination and improvement of theories, whatever their domain of application, and is the basis of all our knowledge of the real nature of the world and the Universe.

2.3 Scientific investigation

The scientific method is nothing other than the basic learning cycle just discussed, but applied in a systematic manner. The power of science as a method of investigation arises from two features it has practised to perfection: firstly, the use of the *analytic method*, i.e. dividing a system into its parts and understanding how the parts work in isolation from each other; and secondly, the systematic application of *quantitative analysis*, used in conjunction with measuring instruments of ever increasing accuracy, so that the regularities underlying nature are formulated as *mathematical laws*. These features have been employed together in formulating theories and devising precise experimental tests of the theories proposed, leading both to the ability to predict to great accuracy the behaviour of simple isolated systems, and also to considerable knowledge about the nature of the constituent parts of matter.

One of the remarkable features that has emerged is the amazing power of mathematics [12,13] in describing the nature and behaviour of matter; it is something of a puzzle as to why this should be so (see Barrow and Davies in [6]). It should be emphasized that this has become particularly apparent in recent times, due to the the dramatic improvements in measurement technology: we can now measure times, distances, masses, and other properties of objects to incredible accuracy, due to enormous improvement in imaging and measurement techniques, often through use of entirely new processes (scanning microscopy, Nuclear Magnetic Resonance imaging, laser interferometry, and so on).

In searching for scientific laws, we are looking for invariant behaviour common to many systems, providing a unifying explanation of different generic and specific cases. We do so by separating out their behaviour into a *universal part*, 'laws of nature' applicable to all similar systems, and *specific information* determining the response of a particular system to those laws, usually in the form of 'initial conditions' and 'boundary conditions' specifying the nature and state of the specific system under investigation. Thus, for example, general laws of motion describe how any falling object moves; the particular place and speed with which a ball is propelled, together with a specification of wind conditions, determines its particular path and pace of motion. Experimental tests function by varying some of the initial conditions of the system, keeping the rest fixed, and then seeing if the response to these new conditions follows the universal pattern described by the laws we suppose are applicable to that situation. So we may release projectiles with various velocities from a tower, and verify if the way they fall complies with our theories of wind-resistance and gravity; the test confirms the theory if they move precisely as predicted (within the experimental error). By carrying out such tests, we have been able successfully to confirm physical laws that do indeed describe accurately the behaviour of a large variety of physical systems and enable us to predict their future behaviour with precision.

Different kinds of science

While the characterisation just given describes the nature of a large part of the natural sciences, including the fundamental sciences of physics and chemistry, it is important to realise there are other forms of 'hard' science with major differences in their practice.

Firstly, there are the purely *logical sciences*, specifically logic itself and most of mathematics, which are not susceptible to experimental test[4] or proof. Rather they are based on pure analysis and examination of the satisfactoriness of that analysis. The other sciences build their theories on the basis of the logical sciences (the analysis of physics is based on mathematics, for example).

Next there are the *natural sciences*, comprising the *analytic sciences*, as outlined above, and the *integrative* (or *synthetic*) sciences: for example ecology, where the emphasis is not on the behaviour of the parts of a system, but rather on the behaviour of a complex system made up of many interacting parts (the behaviour of each of these parts being susceptible to analysis by the analytic method). The *applied sciences*, such as engineering and computer science, fit into this category, and in many cases remarkable success has been achieved; for example we can now use computer simulations to predict the behaviour of aircraft before they have been made, on the basis of Newtonian dynamics and the theory of fluid flow. As in the case of the analytic approach, these understandings of how complex systems function are open to experimental test (provided we can isolate the system adequately from outside interference). One of the main differences from simple systems is that there is usually no way we can test all possible types of initial configurations of a complex system; we can only check its behaviour under a representative sample of initial conditions, and hope that this gives us sufficient insight into its behaviour in the face of all the conditions that will be encountered in reality. (For example we test the computers that control aircraft in a way we believe will adequately reflect the whole range of conditions they will encounter in practice; but we cannot test all conditions that might occur).

Clearly the synthetic sciences build on the analytic sciences, in that as we attain a better understanding of the behaviour of the components of a complex system, we are in a better position in our attempts to comprehend the whole. However they cannot be reduced to the analytic sciences: it is precisely the relations between the parts of a human body that enable the whole to function, and this cannot be understood by examining the parts in isolation. This is why physiology is of necessity an integrative science. Furthermore one should note that while the objects of study in the reductive sciences may often be identical to each other (each proton is identical to every other proton, similarly form electrons and neutrinos), in the case of the integrative sciences there will, despite major similarities, always be some differences (each human body is different from each other one; each ecosystem has its own individuality).

Finally there are the *historical and geographical sciences*, such as geology and astronomy, where we examine the nature, history, and origin of unique systems (a particular mountain range, the Earth, the Local Cluster of galaxies, and so on). Major examples are the investigation of the creation of the Solar System, of the history of continental drift on Earth, and of the evolution of life on Earth. In these cases, we cannot set initial data so as to repeat the situation that occurred in the past: experimental tests in the sense implied above are not possible, because we are concerned with specific events that have only happened once in the Universe. However we can on the one hand look at properties of similar systems or events, hoping they will help us understand this particular one (the issue of course being just how similar or different all these other examples are); and on the other hand we can today make observations of many kinds of data that tell us much about the specific historical event of concern, representing features that would be necessary consequences if our theory is correct. (For example we can search for similarities in the DNA patterns of animals that we believe to be closely related through evolution, or we can compare ages of different fossils as determined by measurements of radioactive decay.) Thus we can try to predict the results of observations that have not yet been made, on the basis of our current best theories about the specific historical event in question (we may predict that various rocks must be similar in South America and South Africa if continental drift is indeed true, and then go out and verify if this is so or not). We may also be able to measure present behaviour that tends to confirm our ideas about the past because it is the same process going on at the present time. (We can for example measure the present rate of change of distances between the continents, and observe mutations presently occurring in population species.)

All of this provides corroborative evidence which may be very convincing, but is still not the same as observing the unique course of events that took place in the past, finding out the effect of altering initial conditions at that time, or carrying out experiments that repeat the same course of events. The historical sciences build on the analytic and integrative sciences in that these give guidelines as to the kinds of behaviour to expect, and put strict limits on the kinds of things that could have happened in the past[5].

Science

Logical science	Natural science		Historical science
	Analytical	**Synthetic**	
Logic	Physics	Ecology	Geology
Mathematics	Chemistry	Astrophysics	Evolution
Statistics	Molecular Biology	Physiology	Astronomy

Table 1:
A classification of the different kinds of sciences, with examples of each listed [Note that applied sciences and social sciences are not shown here.] The Natural Sciences and Historical sciences together comprise the observational sciences.

Table 1 shows this classification of sciences (apart from applied sciences and the social sciences), reflecting the relation of each to testing and confirmation. I use the latter word advisedly; it is not possible to verify any theory in the sense of proving without doubt that it is correct, but we can confirm it by providing more and more evidence that supports its correctness (for example no one seriously doubts that Newton's laws of motion adequately describe how a motorcar moves). The use of Bayesian statistics gives such confirmation a solid logical foundation [14-16]; in this approach we always regard knowledge as incomplete, but with each new bit of positive evidence adding to the previous evidence for believing a theory is true. We can sometimes disprove theories, including historical theories, in a decisive way, when observations clearly contradict some of their predictions. (For example, the finding of a human skull that radioactive dating proved came from the era of the dinosaurs would confound the present theory of evolution.)

The invariant underlying nature

The point that now needs to be made, in the face of much recent writing emphasizing the sociological basis of scientific activity and suggesting that all scientific understanding is therefore culturally bound and relative (see e.g. [9]), is the efficacy of physical laws. Social and cultural issues do indeed play an important part in shaping science, for they help determine what is regarded as an important issue at any time, and therefore help shape what questions are asked by the working scientist; and to some extent they shape the kinds of theories proposed to provide explanation. However the fundamental point is that, provided these questions and theories are then refined and developed appropriately to lead to true scientific tests, the answers obtained (in the logical and natural sciences) are not relative: on the contrary, they reveal some aspect of the working of nature that is universal. It is independent of the time and place where the experiment takes place, and of the cultural and sociological nature of society. So when I release a weight it drops to the ground; water is always made of hydrogen and oxygen; electric waves propagate at the speed of light; and so on[6]. Indeed the functioning of the world around is rigorously determined by the laws of physics and chemistry, whatever we may wish or do, and no matter what our life view may be; *inter alia* these laws determine the functioning of our own bodies and minds, the physiological nature of that functioning also being completely determinate (we breath oxygen; DNA determines our genetic inheritance; calcium and sodium ions are the basis of signal transmission in our nervous systems; and so on).

This is the triumph of natural science. It has been able to locate an underlying reality that is invariant and universal[7], despite the differing social and cultural positions from which different scientists operate. We expect motor cars, television sets, refrigerators to work in a specific way that results from their design; and they do so completely reliably (unless defective) precisely because the laws of physics govern their behaviour in a reliable and repeatable way.

While the implications of the natural sciences may be absolutely firm in many cases, the description we use may not be rigidly specified. Indeed there may be different viewpoints and even different mathematical descriptions that describe the same processes equally well, and give the same results. For example, there are different ways of formulating Newton's theory of gravity: in terms of a potential (the field view), in terms of action at a distance (the force view), or in terms of a variational principle (a Lagrangian or Hamiltonian view). In the case of quantum mechanics, apparently different mathematical descriptions (the Heisenberg, Schrodinger, and Feynman descriptions) have been shown to represent the same physical behaviour: they are equivalent descriptions, although this is far from obvious. In these cases we may prefer one or other of the alternative views of the same phenomenon on cultural or aesthetic grounds; this will in no way alter the solid nature of the predictions obtained from them.

Criteria of choice for theories

The situation is not so clear cut in the case of the historical sciences (as defined above), such as geology, evolutionary theory, archaeology, and astronomy. Here there may well be a cultural or sociological bias influencing the conclusions we derive (as well as the description we use), for in the face of our inability to perform the experiments we would like to carry out, we have to make assumptions determining what kind of theory we regard as reasonable, and the shape of what we find is biased by these assumptions.

Thus a key issue is what are appropriate criteria for a satisfactory theory, that can help us choose amongst possible alternative explanations. The primary candidates are,

* *simplicity* - the Occam's razor idea that one uses the simplest possible theory that can accommodate the facts;

* *beauty* - on the face of it a very subjective criterion, but there is remarkably good consensus about it in many cases;

* *prediction and verifiability* - the ability to confirm the theory by a variety of observations or tests; in particular, verified *predictions of a new kind*

(such as the existence of anti-matter, the bending of light by the sun, or the transformation of matter into energy) provide major support for correctness of a theory. The converse is the Popper criterion that a good theory should be clearly *falsifiable by experimental test*;

* *overall explanatory power and unity of explanation*, in particular congruence with the rest of our current body of knowledge.

These are the key requirements for a 'good' theory; the problem is that in general these criteria will not agree, and differing emphasis on which of them is important may lead to different choice of theory [8]. Nevertheless in many cases we may attain a high degree of certainty because these criteria concur, selecting uniquely as preferred a theory with high explanatory power that also fits into the body of established theory in a satisfying way.

We will take this latter feature as a prime requirement, insisting that theories we propose are as far as possible based on the same principles as the well-established analytic natural sciences, where abundant tests are available, unless there is good reason to believe this is wrong. Then we move to the immediate evidence for a particular historical interpretation of a series of events, also that mass of evidence which supports the surrounding body of theory that is congruent with this interpretation; then this all provides corroborative evidence of the overall scheme of understanding. Given such a consistent scheme of interpretation, there will be many historical situations where we are destined always to remain in doubt about the true course of events, because of the fragmentary nature of the evidence available to us; for example many cases in archaeology, and many of the details of evolutionary history. Thus we will attain various degrees of certainty, according to the quantity and quality of evidence available to us.

Reasonable certainty in historical sciences

In particular cases, there is a vast interlocking array of facts that are explicable in a unitary way if we adopt one particular explanation of a complex phenomenon, but remain a set of disconnected features that have common aspects by pure chance if we do not. By marshalling evidence in this way, we can be virtually certain, for example, that continental drift and evolution did in fact take place, and that carbon dating gives correct estimates of ages of archaeological finds. It is possible to put forward logically consistent alternatives that explain the same historical facts differently, such as the so-called 'creationist' view; the problem is their incongruence with the rest of scientific knowledge, together with the small number of features which they explain. They provide a consistent scheme, but considered as a scientific scheme of interpretation and understanding, it is one of narrow scope and small vision, with low integrative power; later on we will conclude it is also a scheme of narrow scope and understanding when viewed from the philosophical and religious viewpoint.

It is interesting to view this whole discussion from the experience gained in law courts. For the examination of evidence as undertaken there is nothing other than applied historical research; and it is taken most seriously because people's freedom and livelihood hang on the verdict. The crucial point is that in many cases we believe that, despite all the pitfalls, we are indeed able to arrive at a verdict, 'beyond all reasonable doubt'. The cautionary remark is that sometimes such a verdict is wrong.

The upshot is that in the case of the historical sciences, an element of interpretation is implied by any theory, as a matter of necessity; there is no way to avoid this. We minimise this as far as possible by demanding consistency with present day scientific theory and understanding, together with the requirement of broad explanatory power. This will in many cases lead to a unique interpretation of specific events that fits in with present scientific theory in a mutually reinforcing way, so that an unrivalled interpretation is both attainable and in effect required: for example, the evolutionary principle not merely becomes an explanation of past events but becomes a central feature and profound organising principle of biological theory also explaining many events occurring today.

There is always a chance that at some later time a new interpretation will be found to be better, in the light of further knowledge that becomes available; however this is always the case for all forms of knowledge, for as has been emphasized above knowledge can never be regarded as absolute.

2.4 A broad approach

Given this view of the 'hard' sciences, we can apply the same ideas more generally in our broader search for understanding. We will in general be involved in the same struggle to determine generic patterns of behaviour that apply to particular situations. Where the systems are very complex and the number of interacting factors is very large (as is the case in the social sciences), attaining universal laws - that is, descriptions of behaviour that always apply - is difficult, but not impossible if the level of generalisation is appropriate[9] For example the computer statement 'Garbage in, Garbage out' is of universal applicability to all information systems. However such descriptions may not be attainable at a desired level of detail. At a more detailed level, by and large we attain only broad generalisations that show how things generally work. Quantitative information may be helpful or even essential in attaining these understandings, but there will be some areas where it may be irrelevant, particularly those involving value judgements or the assignation of meaning; and a central feature will be the way we handle the interpretation and understanding of specific individual events. In any case we will need to use general guiding principles, such as those suggested above, to choose between different possible patterns of explanation; thus the most crucial choices we make are in deciding on those guiding principles.

Whatever the field of concern, the basic learning cycle discussed above is the way we obtain and extend our understanding. Employing this method consciously will increase its effectiveness. This demands

1. *an openness to possibilities*, the readiness to search out grains of truth that may be hidden in alternative viewpoints and a willingness to test them for possible validity. Thus a key question one can ask oneself is,

> What am I prepared to question, and what am I not prepared to question?

The answer lays down the fundamental parameters within which one is willing to learn. Those issues that one is unprepared to question are those where one has chosen to proceed on the basis of preconception and dogmatic assertion, rather then reflective investigation. The probability that the conclusions will be correct is very small (why should reality conform to unreflective preconceptions, when in many cases we know these are wrong?) Examples of issues where such assertions are frequently made are, the correctness or not of evolution; the existence or not of God;

the reality or not of psychic phenomena; the viability and rightness of various economic systems; the ethics of abortion (the last two raising issues of value judgements that we return to later). If the mind is closed on some point, that should be stated at the beginning, so that we all know that rational discussion on that issue is not possible. The adoption of ideologies of various kinds is a common way of protecting oneself from questioning what one holds dear. The aim of the approach suggested here is that one will try to avoid having a closed mind on any issue whatever. This means in particular, emphasizing freedom of information and freedom to support ideas other than the current dogmatism. The basis of progress and understanding is an atmosphere conducive to enquiry [11].

Developing a consistent theory based on this approach means integrating the possibilities considered into a coherent and logical scheme; our logical and creative skills come into play as we fashion a satisfactory whole. A fundamental point here is

2. *identifying the important issues and concepts*, naming them, and characterising them. We can then show how they relate to each other and to the extant wider body of theory and knowledge. The question then is:

> What are the key elements underlying the important features of interest?

The point is not only to state what is fundamental and what is less important (from a causal viewpoint), but also to consider what may have have been left out, perhaps because it is so obvious that it is taken for granted and therefore not taken into consideration.

Having set up a theory (or set of competing theories) our task is to separate the wheat from the chaff by suitable testing. The key point here is

3. *determining valid methods of testing theories and of assessing the results of the tests.* This will vary greatly from area to area. We have to ask:

> What is acceptable data for this area and how is its quality assessed?

If some feature regarded as key by some is not admitted as data at all by others, agreement cannot be reached. However often the disagreement is not over the general admissibility of a class of data, but the quality of specific data. These problems apply particularly to historic records on the one hand, and to personal experience on the other. We are obliged to take personal experience seriously as data, for it is the central feature of each person's life; but we are very aware of the ability of our senses and our mind to deceive us about what is really going on.

Thus it is essential to develop adequate ways of assessing which personal experiences (for example, claimed psychic or religious experiences as well as individual observation of experimental apparatus) can be regarded as valid, and which as hallucination or self-deceptions, if we are to have a sound coherent basis for understanding in the broad sense sought here. That this may not be impossible, if approached with care and circumspection, is suggested by the experiences of the law courts on the one hand, and of some psychotherapists on the other, although the latter suggestion may be controversial. What is beyond contention is that many people believe psychotherapy gives unrivalled understanding of unconscious mental processes that must be taken seriously if we wish to understand ourselves and our nature in depth. This in itself is significant data that can be taken into account.

I will return to these issues in Chapter 8. At present two comments are in order as regarding the nature of evidence.

Firstly, we must realise that in practice serious counter-evidence to a theory may not, by itself, lead to the theory being dropped, but instead to the investigation of various *perturbing influences* which have not been taken into account so far (the wind, a temperature gradient, an electric field, that may have affected the path of a falling object; the chemicals are not pure; an unknown person entered the room and interfered with the murder scene before the deed was discovered). The point here is that even if we are investigating identical systems (say electrons), each experiment is in fact an individual experiment that is in detail different from every other experiment. We have to take this difference into account in interpreting the experiment.

Secondly, whenever observations are made in support of a theory, we should always be aware not only of possible distortion of the data but also of the various possible *selection effects* that might be in action, distorting the appearances of what happened by effectively preventing a whole segment of data getting through. This happens not only in science (for example, they determine what galaxies we detect in photographic plates of distant regions of the sky, because many are too faint to be detected) but also in such cases as newspaper reporting and the way messages are conveyed to those in charge of bureaucratic or military structures. Thus a particular question we need to always ask is: What is the data we are not taking into account? What is the information that is not getting through? Furthermore, as emphasized in the famous Sherlock Holmes comment on the strange incident of the dog in the night (the dog did not bark), the absence of a signal may convey vital information which we need in order to understand some situation.

4. Finally, having analysed the competing theories and considered the evidence for and against them, we need to use our *chosen criteria for good theories* to choose between them, asking

> How good is each theory relative to alternatives, in terms of our criteria?

The rest of the book gives examples of how this may be done in various important areas.

2.5 The Essential Need

Overall, the essential feature advocated here, in line with one of the major movements in methodology of the last half-century, is that our concern ultimately must be an emphasis on process rather than the particular presently available data and knowledge. For if the process is right, then in due course errors in our understanding will be corrected.

> *Thus the basic need is a process of learning that courageously investigates reality and formulates unifying hypotheses, but also continually checks for incongruities and problems, and corrects the errors found.*

This chapter has briefly set out the nature of such a process. Its implications will be developed in the following chapters, considering firstly the nature of the physical Universe, and then the best way to develop a coherent broader perspective.

PART TWO

THE SCIENTIFIC UNDERSTANDING

CHAPTER THREE

THE PHYSICAL FUNDAMENTALS

This chapter considers the basic understandings of physics and chemistry about the particles that make up the physical world, the forces that control their behaviour and the general principles by which these forces operate. This provides the foundations on which we understand the functioning of complex systems and astrophysical objects discussed in the following chapters.

3.1 Components and forces

The basic question, What are things made of ?, has been answered in a comprehensive manner, as follows: matter has a hierarchical structure[1.] The macroscopic objects (i.e. objects of everyday size) we see around us are composed of *molecules of* very small size; these are the basic units of chemical compounds . One should realise the huge numbers involved: one cubic centimetre of water contains about 10^{24} molecules (that is, about a million million million million molecules). Each molecule in turn is composed of atoms, the basic units of chemical elements, which are the simplest identifiable chemical substances (such as hydrogen, H, carbon, C, and oxygen, O); there are 92 naturally occurring elements [18], each with a specific atomic structure and characteristic physical and chemical properties. Each kind of molecule is formed from atoms bonded together in specific combinations. For example water is H_2O (i.e. a water molecule consists of two hydrogen atoms and one oxygen atom), methane is CH_4 (a methane molecule consists of one carbon atom and four hydrogen atoms), and so on. Atoms are composed of a positively charge *nucleus* at the centre, much smaller than the atom, surrounded by negatively charged *electrons*, much smaller than the nucleus. A nucleus is composed of positively charged *protons* and uncharged *neutrons,* together called nucleons. The simplest theory is that of ordinary hydrogen consisting of just one proton; the most complex natural element is Uranium, with a nucleus containing 92 protons and either 143 or 146 neutrons (elements occur as different *isotopes*, with the same number of protons but different numbers of neutrons, and hence different masses).

Protons and neutrons are made of yet smaller particles called *quarks*. There are 6 types of quarks, each occurring in three different internal states (called 'colour' states, but this is not related to ordinary colours; rather these states are really like having different charges); and there is an anti-quark for each quark. Each nucleon is made of three quarks. This is the limit of resolution obtainable at present; quarks and electrons may be "elementary" (i.e. indivisible) particles, or may be composed of even smaller components. We have various theories about this, but do not have the evidence needed to make a decision [19,20].

One of the aims of science is to pursue this issue to its limit, that is, to try to determine what are the 'ultimate' constituents of matter. However the description already given is sufficient for practical purposes, in the sense that the behaviour of matter under ordinary conditions is explained by its molecular and atomic structure, together with a knowledge of the forces acting on the particles making up atoms. The chemical properties of an element are determined by the arrangement of electrons in orbits around the nucleus, their number being equal to the number of protons in the nucleus (so that the total electric charge is zero).

One of the great triumphs of physics is that it is able to explain not only *the periodic table*, the systematic characterisation of the properties of all the elements, but also the way elements combine to form compounds when chemical interactions take place [21,22], which is highly non-obvious (for example the silver metal Sodium and the yellow gas Chlorine, both poisonous, combine to make ordinary table salt - forming white crystals that are so desirable as a condiment that empires have been built on their trade). By physical and chemical analysis, we can determine the atomic nature of the molecules that are the basic structure of the objects we see around us, and even (despite their immense complexity) of those that are the materials out of which our bodies are made [23,24]. These *organic molecules* (the molecules out of which living things are formed) are often vastly larger and more complex than the inorganic molecules out of which ordinary materials are made (they can contain hundreds of millions of atoms).

Overall we have then a hierarchical structuring of matter at small scales (probably with smaller scales still to be resolved someday):

Large organic molecules
Small Molecules
Atoms
Nucleons and nuclei
Quarks and electrons

Table 2: Hierarchical nature of microscopic structure.

Fundamental forces

The behaviour of the molecules, atoms, and particles is governed by the forces that act on them, which are described by physics [25-28]. There are four fundamental forces, and nowadays we understand these forces as being mediated by particles which convey energy and momentum, and bind matter together [19]. The forces are,

* *the strong force*, which binds nuclei together;

* *the weak force*, which controls radioactive decay of a neutron (into a proton, an electron, and a neutrino, the latter being an uncharged particle of very low mass, that conveys energy and momentum when interactions take place);

* *the electromagnetic force*, which governs the interaction of charged particles, mediated by photons (particles of light, by which we see the world around us); and

* *the gravitational force*, which governs the structure and motion of astronomical objects (as well as making objects fall on the Earth).

The strong and weak forces are short-range interactions, and so can only be effective at microscopic scales. The strength of the electromagnetic force between two bodies depends on their electric charge (which can be positive or negative), while the strength of the gravitational force depends on their masses (which can only be positive); both are long-range interactions, the strength of the force exerted being proportional to the inverse square of the distance between the interacting objects. It is these two forces that dominate our everyday lives; for example we use the electromagnetic force in the functioning of electric motors, telephones, computers, radios, and television sets, while it also governs the basic nature of chemical interactions, and so of the functioning of living systems, including the human body [22,24].

3.2 Overriding principles

The way these forces operate is governed by three over-riding principles: the principle of relativity, the principles of quantum theory, and conservation of energy and momentum. Their practical implications are often determined by the statistical behaviour of large numbers of particles.

Relativity Theory

Firstly, all forces obey the *principle of relativity*, which states that the laws of physics are the same irrespective of the observer's state of motion. This apparently innocuous statement has profound implications for the nature of space and time [28,29-31]. When combined with the remarkable experimental result that nothing can move faster than the speed of light (about 300,000 km/sec), it implies that the speed of light is measured to be independent of the motion of an observer. This then leads to the phenomena of length contraction, time dilation, and relativity of simultaneity for relatively moving observers, quite contrary to commonsense understanding. *Special Relativity Theory* unites these phenomena in a description that emphasizes the relativity of most physical measurements, in that their outcome depends on the relative motion of the observers making the measurements. Yet the theory also emphasizes the existence of an invariant underlying reality that can be determined by those observers, namely a spacetime combining in one unity features we normally associate separately with space and with time [31].

Relativity theory also transforms our understanding of the relation be-tween matter and energy: they are equivalent, in that they can be trans-formed into each other (an essential consequence of the theory of relativ-ity, expressed in the famous equation $E = mc^2$ relating the energy E and mass m of an object through the speed of light c). This is the foundation of the use of nuclear power both by people (in nuclear reactors and weapons) and by nature (in the stars and the sun; nuclear reactions provide the power source enabling them to shine). However the other major implications of relativity theory are not readily discernible in everyday life: they become important only when objects move at nearly the speed of light, much faster than the fastest aircraft or rockets ever built.

Our understanding of gravity is also revolutionised by the relativity principle, when this is combined with Galileo's fundamental discovery that all bodies in a vacuum fall equally fast, irrespective of their nature. It follows that a uniform gravitational field can be transformed away if one changes to an appropriately accelerating reference frame (gravity appar-ently vanishes if one is in free fall). Consequently gravitational and inertial forces are essentially the same (this is Einstein's *Principle of Equiva-lence*), so gravity is not a force like other forces, but is essentially a consequence of space-time curvature [29,31]. The amount of this curva-ture is determined by the amount of matter present, through the Einstein Field Equations for gravity. As the geometry of space-time is determined by the matter in it, this geometry is no longer a fixed and given thing, as had been previously assumed, but rather is dynamic and changing, controlled by physical laws. This understanding (formalized in Einstein's General Relativity Theory, the modern theory of gravitation) leads to prediction of phenomena such as the bending of light by a gravitational field, and the existence of black holes (discussed later).

Quantum Mechanics

Secondly, all matter conforms to the *principles of quantum theory*, one of the most surprising and counter-intuitive discoveries of the century [19,32-34]. In brief, energy is exchanged only in discrete units called quanta[2], which can be thought of either as particles or as waves; indeed all matter behaves both as particles and as waves, when viewed on the smallest scales. As a consequence there is a basic uncertainty in our ability to know the position and behaviour of fundamental particles, and in our ability to predict the way microscopic systems behave. This is an essential feature of nature. It implies that we cannot predict with certainty what systems of atomic or sub-atomic scale will do; for example we cannot tell when a radioactive atom will decay, or which path a particle will take. This uncertainty is a profound and apparently irremovable feature at the basis of sub-atomic physical behaviour [19].

When combined with relativity theory, quantum theory predicts the existence of *anti-particles* for every kind of particle (with the same mass but opposite charge.) For example the positively-charged positron is the anti-particle of the electron. A particle can combine with its antiparticle to become pure energy, and the opposite can take place also, i.e. photons of very high energy can collide with each other to produce particle-antiparticle pairs ('pair production'), where previously there was nothing but energy. We do not encounter anti-particles in ordinary life on earth, but they can easily be created in high-energy colliders.

Quantum theory underlies the stability of matter and (through the *Pauli exclusion principle*, which prohibits more than one particle of matter of a particular type filling the same quantum state) is the basis of understanding the periodic table of properties of the elements. Thus it is the foundation of extremely important properties of everyday matter. However the quantum nature of that matter is not easy to discern directly, because the size of quanta (the discrete units in which energy is exchanged) is so small. For example light appears to us to be of a continuous rather than of a particle nature; however modern detectors can detect individual photons (i.e. individual light particles) coming to us from very distant galaxies.

Quantum theory explains an important property of the light emitted by chemical elements at high temperature: the existence of *spectral lines*. When light from a star is separated out into its different colours by a prism, one obtains its spectrum. (The light is spread out into its different colours, enabling us to see how its intensity varies with colour; as the colour is determined by the wavelength, one can measure the variation of its intensity with wavelength). The significant feature then is that very sharp lines are seen in the spectrum, these lines being characteristic of the material emitting the light. For example Hydrogen creates highly ordered series of lines known as the Lyman and Balmer series. Quantum theory gives a complete explanation of how this occurs (and correctly predicts the wavelengths of the lines). This feature is very helpful to us in understanding the nature of distant objects; for example we can analyse what chemical elements are present in distant stars by examining their spectra in detail.

Conservation of energy and momentum

Thirdly, all interactions obey the laws of *conservation of energy and momentum* [25-28]. Because mass and energy are equivalent, conservation of energy implies that mass is conserved also. However different forms of matter and energy can be transformed into each other under suitable conditions, provided various conservation laws are obeyed which determine what are and what are not allowed transformations. Thus many chemical transformations between different elements and compounds can take place at room temperature; always the number of each type of atom involved in an interaction is conserved (e.g. when two atoms of hydrogen and an atom of oxygen combine to form one molecule of water, the number of hydrogen atoms - two - and of oxygen atoms - one - is the same before and after the reaction; the difference is that before they were free but after they are bound together). Examples abound: burning wood, cooking, brewing beer, digesting food; in all cases we can convert different molecules into each other, but we cannot change the number of each kind of atom involved in the interaction. On the other hand when nuclear reactions take place, involving very high temperatures, the specific atomic nature of matter can be altered: hydrogen can be transformed to helium, or nitrogen to oxygen, indeed in principle lead can be transmuted to gold (but the cost of doing so is very high!)

When such reactions take place, the numbers of the constituent particles (protons and neutrons) making up the atomic nuclei involved are always conserved. However these particles in turn can be transmuted into each other when weak interactions take place; but then the total *electric charge* is conserved, determining how many more positively charged particles there are than negatively charged particles (the number being negative if there are more negative particles than positive ones). Similarly other quantities that are conserved under all but the most extreme conditions (the *baryon number*, determining the number of protons and neutrons, and the *lepton number*, determining the number of electrons and neutrinos), guarantee that while transmutations take place, an essential character of the matter involved is unchanging [19].

Thus a variety of interactions allow different transformations of the form of matter, depending on the conditions (there is usually a threshold temperature below which reactions are not possible, but above which they readily take place [22], as we experience for example in lighting a fire). At the temperatures prevailing on earth, for all practical purposes the quantities of different chemical elements available to us is fixed and finite. This conservation law is the basic factor underlying the significance of scarce resources.

Regardless of what conditions are in any interaction, always the total energy involved is the same before and after, if we do the accounting correctly, in particular taking heat energy into account (and considering mass changes, if nuclear reactions are involved). This is true on the microscopic scale and on the macroscopic scale; indeed conservation of energy is the *First Law of Thermodynamics*, one of the most basic limits on what is and is not possible in the real world [22,25-28,35]. Thus chemical energy stored in a battery can provide electricity to run a motor or to light a flashlight; burning of coal turns chemical energy stored in the coal into heat energy, used to warm a house or cook food; nuclear energy stored in uranium can be converted into electrical energy in a power station; the energy stored in food can be turned into available internal energy in the body, and then into gravitational potential energy when a mountain is climbed; that potential energy can be converted into the energy of motion (*kinetic energy*) if the climber falls; the kinetic energy in water or wind can be converted to rotational energy by a windmill, and then used used to grind corn; and so on. Many forms of transformation of energy are possible; in every case the total energy used is equal to the energy provided from some store. Thus what we can achieve is limited by the energy stores at our disposal: we cannot attain any aim that requires more energy than we have available.

The same is true for total momentum [26-28]. Conservation of momentum of a single body (when no unbalanced force acts on it) ensures it will keep moving in an undeviating direction at constant speed. If some unbalanced force (air resistance or gravity, for example) acts, its momentum will change according to the magnitude and direction of the total resultant force acting, the consequent change of momentum determining how it will move. We can therefore alter motion by exerting forces appropriately, controlling the movement of motorcars and aircraft, of cricket balls and falling rocks, of the human body as a whole and of each part of the body. In this way the dynamics of daily life is controlled by the laws of motion.

Entropy and the Second Law

However there appears to be a paradox about energy conservation. If I slide a book across a table, it slows down and stops (because of the friction exerted on the book by the table top). The kinetic energy (the energy of motion) appears just to decay away: where has it gone? It appears as if energy is not conserved. The answer is that it goes into heating up the table: the table top will be a little bit warmer after the book has slid by than before, because it has absorbed energy in the form of heat. Thus total energy is conserved. However the key point is that *the transformation is irreversible because the heat energy is irrecoverable*: we cannot re-collect it and turn it back into kinetic energy, thus setting the book in motion again. [34] (We can set the book in motion again if we expend extra energy by giving it another push, but not by re-using the energy which is present in the surface of the table as heat.) This is an example of the *Second Law of Thermodynamics*: there is a *growth of entropy* (i.e. an increase in unusable forms of energy, commonly characterised as 'disorder') that inexorably takes place whenever interactions occur between macroscopic objects. The consequence is that energy is continually being degraded into unusable heat energy, so that although the total of energy is constant, the amount of usable energy in any isolated system is continually declining [22,25,28,34,36].

This is one of the most fundamental restrictions on what is achievable in practice in the real world: it implies we can never achieve the theoretical ideal indicated by the First Law of Thermodynamics. Thus the design of motorcars and steam engines, of aircraft and refrigerators should always aim to minimise this loss of useful energy, which is inevitable. The same principle of decay is true for scarce materials: for example although the quantity of iron in the world is constant (ignoring the small change that might take place in nuclear reactors), chemical processes such as rusting continually degrade iron from readily usable forms to a form where it is not easily accessible for use. Similarly any human activity creates waste products that have to be disposed of; they are 'waste' because the useful input form (for example, coal for the fireplace) has been reduced to a form that is useless for our present purposes (for example, ash); it may be possible to convert some of it back to a new usable form, but there will be an energy cost associated with so doing. Much of the environmental battle is associated with the disposal of waste products, or finding alternative uses for them (the ash, when appropriately treated, can be used as fertiliser). Among the consequence of this uniform trend of physical systems to disorder is that we need to value highly ordered deposits of rare resources. For once they are used and dispersed, although the elements are still there, for practical purposes they are gone (the cost of recovering them is too high to make it feasible).

The laws of large numbers

The Second Law of Thermodynamics is a statistical law: in principle it could be violated in highly exceptional circumstances, if everything conspired just right. In practice this will never occur in the case of ordinary size objects, for millions of millions of particles would have to conspire exactly to make this happen (all the scattered molecules of a broken glass of water on the floor could in principle precisely reverse their motion and make the glass come together again, [34]; no one has seen that happen.)

This is an example of an important feature of physics: when very large numbers of particles are involved in an interaction, *statistical mechanics* comes into play as an organising principle, based simply on the large numbers of particles that are interacting. It enables predictions of their properties to be made, that are to a considerable degree independent of the details of the particles involved and of the forces in action. One important example is the 'perfect gas laws' describing the behaviour of most gases, largely independent of their composition; these result simply from the effects of myriads of collisions between the gas molecules, occurring all the time. Another example is that when matter and radiation are in equilibrium with each other in a closed box with an absorbent surface, then independent of the nature of the matter, the radiation will become Black Body radiation, i.e. it will have a characteristic sharply peaked spectrum (very similar to that of the Sun), with the wavelength of the peak uniquely determined by the temperature of the matter. One of the triumphs of quantum mechanics is the detailed theoretical explanation of this spectrum. Another is the explanation of the specific heats of matter, i.e. the differing abilities of different types of matter to absorb heat. In working this out, it is an essential feature that the fundamental particles of any particular type are all completely identical to each other (there is no distinguishing feature marking any electron as an individual particle: they are all precisely identical to each other in their structure and behaviour) [19]. By contrast, macroscopic objects are all individual (each grain of sand is different from each other one).

Closely related to these statistical properties is the fact that while the quantum uncertainty principle prevents precise prediction of the behaviour of sub-atomic systems, the laws of physics and chemistry applied to macroscopic objects (comprised of a very large number of molecules) give extremely accurate prediction of their behaviour and motion; for example one of the triumphs of science is the prediction of the motions of the planets with incredible accuracy. It is this kind of ability to make quantitative predictions that underlies the advances of modern technology.

3.3 Exotic conditions

When conditions are very different from those of everyday life on earth (for example, at the centre of the sun) the kinds of interaction taking place will be very different from those common on the Earth; and even the nature of the forces acting may change; (under the extreme conditions in the early Universe, forces that appear separate to us may become unified into a kind of super-force). Associated with these extreme conditions will be the appearance in abundance of all kinds of exotic particles that only occur very rarely on Earth (indeed we have to spend vast sums of money building supercolliders in order to detect them). We are still trying to understand the fundamental properties of matter that will determine its behaviour under such extreme circumstances. We have various theories for this, often based on the search for underlying symmetries that order the nature of matter [37,38]; while some are plausible, they are all rather speculative.

A particular feature of the extreme conditions in the early universe, is that understanding them requires a satisfactory theory of *quantum gravity*, i.e. a description that unites the properties of General Relativity (Einstein's theory of Gravitation) and of quantum theory. We do not yet have at hand a satisfactory theory of this kind; Superstring Theory [38] (which represents elementary particles as string-like structures rather than point particles) is a prime candidate, but far from the only one. As we shall see later, this is one of the fundamental sources of uncertainty about the nature of the origin of the Universe.

3.4 Physical reality

We have made vast strides in understanding the nature of the material universe:

> Matter has a hierarchical structure, being ultimately composed of 'elementary particles' (quarks and electrons), interacting with each other through four fundamental forces. The nature of these particles and interactions explains the behaviour of the material world we see around us.

I am not claiming that we understand all the details of how the fundamental particles and forces control the behaviour of the atoms and molecules that are the elementary stuff of every day life; rather we comprehend it in broad terms, and can back up that broad understanding with many detailed calculations that predict quantitatively what will happen in particular circumstances.

CHAPTER FOUR

THE STRUCTURES AROUND US

This chapter considers the way complex hierarchical structures are built up from the basic physical components described in the last chapter. The ability of fundamental particles to form these coherent complex structures is the basis of life.

The basic laws of behaviour and composition of matter determine the structure and nature of the components that make up all natural and designed objects around us. Thus they fundamentally determine what is possible and what is not: all these interactions must obey the principles explained in the previous chapter, in particular the First and Second Laws of Thermodynamics (i.e. conservation of energy and increase of entropy). The question is, how the fundamental laws just discussed permit and enable the behaviour of complex object.

The differing properties of solids, liquids, and gases follow from those of their fundamental constituents [18,28]. This in turn determines their possible combination to serve complex purposes. For example, atoms of carbon, nitrogen, oxygen and other elements form organic molecules in fibres that are used to make cloth, which in turn is fashioned into clothes of many kinds; silicon and other elements are fashioned to form transistors that are combined in printed circuits to form the memory circuits and central processing units of computers; an enormous variety of organic molecules (based on carbon) form proteins, enzymes, lipids, and so on, together forming the immensely complex structure of living cells, which combine together to form the tissues out of which the organs and limbs of living beings are made. The forces between the component parts and the way in which energy is transformed between parts of each of these systems, and so the resulting way they achieve the purpose for which they are designed [1] is determined by the laws of physics and chemistry, which govern the interactions between the fundamental components of any complex system. Thus they govern the behaviour of all the objects and beings we encounter in everyday life: physics governs how the eye sees, how aircraft fly, how we grow and die [39-45].

The great achievement in both natural complex systems and human-made ones, is *reliable operation*. This is true in engineering systems such as cars and aircraft, television sets and computers, and so on. It is equally true in living systems which by and large grow, function, and heal themselves with a perfection of design and operation that becomes more and more astonishing the more one studies the complexity of the interlocking mechanisms at work. It is true that in both cases (engineering and natural systems) there are malfunctions and breakdowns, particularly as they age and the parts start to wear out. But even this language emphasizes the extraordinary achievement: we expect them to function properly, and regard it as a breakdown of what should occur when this is not so. They function as expected provided the component parts are not defective, and have been connected up correctly. However they have a finite lifetime. It has not been found possible to design complex systems that will function for ever. At a fundamental level this is because of the second law of thermodynamics, but one can also argue that death plays a fundamentally necessary role in complex biological systems: it is a required part of the natural order that the old generations give way to the new, life arising from death. Whatever the truth of this, it is not surprising that in systems of such complexity problems eventually arise. The present challenge is to understand how such systems can function at all.

4.1 The major principles

We are facing here the problem of synthesis: combining parts together to form a complex system that functions coherently. There are three basic principles at work in such systems: they are organised hierarchically; their functioning depends on organised information and information flows; and (assuming the system is stable) these flows are arranged to implement the process of feedback control [46-48].

Hierarchical order

The first point is that *order is based on the hierarchical nature of complex systems, implemented as a hierarchy of structural levels.* A motor car is made up of a set of parts (an engine, a gear box, a suspension, a chassis) each of which is made of smaller parts (the engine is made of a camshaft, four pistons, a cylinder block, and so on; the electric system is made of a generator, a battery, a wiring system, a regulator, etc.); each of these in turn is made of its components (the generator is made of a rotor, a stator, ball bearings, etc.). We can readily understand the functioning of the car through understanding this hierarchical design. The success of modern computers is based on the strict hierarchical design of both the hardware (the computer itself) and the software (the programs that run on it). Thousands of transistors, resistors and capacitors in printed circuits make up the central processing unit, the memory units, the display controller, the disk controller, and so on. These function together as a useful computer because the programming languages are arranged in a hierarchical way, the application programmes being written in high-level languages (relatively easily understood) which are based on assembly language (more difficult to read), which in turn is based on machine code (which is very hard to use for direct programming).

In each case one assembles the next layer of structure out of components whose behaviour can be understood separately; these are then connected together in a way that embodies the structural design of the system. One does not need to know the interior structure of each component in order to use it, but only has to understand it as a 'black box'. i.e. a component with unknown inner workings but predictable exterior behaviour. For example I do not need to know all about the internal structure of a battery in order to know how to use one in a motor car. Each component has a specific purpose, fulfilled if it has predictable behaviour as desired, and so is designed to have precisely this right behaviour. It fulfils this purpose if it is 'wired in' in the right way at the right place (the battery needs to be connected in to the electrical system, not to the radiator). In this way we can build immensely complex objects such as a jumbo jet: every single part of such an aircraft has been specifically designed by someone for a particular purpose, and all the thousands of parts are integrated to work as a whole.

The function of each part can be identified and encoded in a suitable name (the wing, the engines, the undercarriage, the instrument panel, and so on); the key to understanding the whole is to focus successively on each level of design and to understand it at that level, starting at the broadest level (the system as a whole) and working down to the level of detail needed. To follow it all the way down one would need to understand the behaviour of the fundamental particles comprising the metal in the cylinder block, but in practice that level of detail is not needed: we simply need to understand that metal is a material with certain properties which we can take as given (unless we are working in metallurgy, in which case it is precisely the question of what makes some metals hard and some soft that interests us).

This principle of hierarchical organisation applies in particular to living systems [49-53]. A complex animal is made of many billions of cells, which together form tissues of various types, these together forming limbs and physiological systems, the whole forming the living being. As we consider each level of structure from the smallest to the largest, we come across the phenomenon of *emergent order* arising from the functional integration of the parts: ''With each step up in the hierarchy of biological order, novel properties emerge that were not present at the simpler levels of organisation'' (Campbell, [49] p.2). Thus ''we cannot fully understand a higher level of order by breaking it down into its parts'' (Campbell, [49] p.3), precisely because doing so destroys the integration that enables it to function as a whole. The same is true for the computer, the jumbo jet, and so on (we will not understand how a computer operates by studying a list of all its parts, even if we are given a description of what each one does, for they could be connected together in many different ways).

Information flows

The second point is that this hierarchical structuring is based on information which it embodies in material form, (the shape of a bird's wing or an aircraft wing is based on information as to how air flows); and indeed *information flows are the basis of organisation*. This is true both in terms of the design of the system (for example plans convey information from the designer to the builder; books inform the designer of the best design strategies; computer programmes written by one group enable design of an aircraft by another), and in the fact that the functioning of any complex system takes place on the basis of flows of information within the system.

For example a motor car is controlled by the driver continually sending signals to the engine (via the accelerator), to the wheels (via the steering wheel), to the brakes (via the brake pedal). A computer is controlled by information stored in a controlling programme; the computer reads the programme and carries out the instructions it finds there; it may also receive instructions from the operator, typing in new information which is used as programme execution proceeds. Of course the programme itself is initially typed in by an operator before it is stored for use. The same is generally true in society: myriads of information flowing through newspapers, telephones, faxes and TV keep society functioning. Central to all complex organisation are structures that obtain, code, store, diffuse, and utilise information.

Most important of all, the information that shapes living systems is incorporated in the DNA molecules that are at the centre of each cell in an organism [23,24,49]. This information, stored in terms of the genetic code and copied to every cell by intricate molecular mechanisms, is read by other molecules as the organism develops, and used to create the hierarchical structure of the functioning animal [53]. In this way an algorithm for the construction of the animal is encoded in DNA molecules, and passed on (in slightly altered form) from generation to generation. Some of this information structures the brain and provides the basis of our own information processing, utilising ways of storing and processing information (possibly by neural networks, or conceivably using holographic techniques) that form the basis of how the mind and consciousness work.

Feedback control

The third point in the implementation of complex systems is *feedback control*. To see why it is necessary, consider a computer operating in a strictly mechanistic pre-programmed fashion. It will (unless some error occurs) do precisely what it is programmed to do, for it is an essentially reliable machine; so that for example we place in its care the safety of passengers in an aircraft, and the keeping of accounts in a bank or building society. However this totally pre-programmed mechanistic approach is also the reason why present day computers are often so stupid, being unable to correct the simplest errors. If a small error is made in a command entered by the operator, the computer may do something quite different from what was intended, or (more often) will refuse to do anything at all (the screen will say ''Command Error'', or something similar); and if the programme itself is wrongly written, the computer will not do the right thing even if the right data is entered. The problem is that the computer proceeds in a purely mechanical way, assuming that the programme is correct and the resulting actions are exactly as planned; but this may not be so. The consequence is that, for example, an aircraft may crash into a hill even if it is computer controlled if the positional information it is relying on is in error.

The missing element is feedback control, which underlies the functioning of all purposive systems. The essential principle introduced is *continual monitoring of what is happening and taking corrective action if things go wrong*. This can be illustrated by considering how you control the temperature of water in a shower. Essentially, you choose a desired temperature T (say, 28 degrees), and compare it with the actual temperature T_I of the water in the shower (to do so you put a hand in and feel how hot the water is). If they are the same, no action is required; however it may be that the water is too hot (T_I is greater than T, say 30 degrees). Then you turn the hot water tap down by an amount you believe is sufficient to cool the water to the right temperature, and (after a minute) feel it to see the result. If it is now too cool, you turn the tap the other way to make the water hotter. By repeating this process several times, you obtain the right temperature.

In essence we compare the actual situation with the desired situation, and if they differ, an error message is generated ('The water is too hot'); this is the information used to actuate a controller (the tap) by an amount estimated to be just right to correct the situation (the corrective information is ''fed back'' to the controller, hence the name *feedback control* . This process is repeated until the right result is obtained. A similar process occurs as we steer a motor car down the road. We observe the direction in which we are heading, compare it with the direction we want to go, and if they differ, we use the amount and direction of this difference (the error signal) to determine how to turn the wheel (the corrective action to make the real situation accord with our wishes). As we drive we continually repeat this process; and we do the same with the speed, monitoring the real speed to see if it is within acceptable limits or not.

This fundamental process is the way that we attain goals in a purposive way. Engineering systems routinely use it to ensure successful accomplishment of their task (for example, in the control of speed of an engine or of the temperature of water). The same process underlies biological purposive action and functioning; an impressive example being the temperature controls in the human body which are so accurate that we can detect ill-health by measuring deviations of our temperature by only one degree from the expected value. An immensely complex example is the immune system that makes our bodies resistant to attack by a host of hostile micro-organisms; and one of the most impressive is the molecular mechanism by which DNA is copied. One strand of DNA is used as a template to make a new strand, the series of four bases that form the genetic code (occurring in complementary pairs) being read to form a complementary strand (containing the identical information). Once in every 10,000 times an error occurs: the pairing of bases is done wrongly. However the DNA polymerase that does the assembly checks the copying for correctness; if a wrong base has been added, it is removed and replaced with the correct one, before the next one is added. After this check has been performed, there is an error only once every billion times.

Feedback control is fundamental, not only to natural and engineering systems but also to human society. It underlies the successful running both of organisations [11, 46, 47] and of personal life [54] and is therefore essential to both basic welfare and adequate quality of life [55, 56]. It is also the underlying principle of the basic learning cycle (see [11]) and of scientific discovery. Whenever feedback mechanisms are missing or ineffective, sooner or later things are bound to go wrong, because the basic mechanism needed to correct errors is missing.

4.2 Ecosystems

The totality of living beings in an area forms a complex interacting network called an *ecosystem* [58,59]; the *biosphere* is the name given to the ecosystem of all living things on Earth. Two new principles emerge here: resource cycles, and interdependence.

Resource cycles

Because material resources are conserved and are limited, any such system depends ultimately on recycling the resources at its disposal. Thus in any ecosystem there are carbon, nitrogen, oxygen and water cycles that maintain the whole in operation; each of these materials spends some time free in the environment, then is taken up by living organisms and incorporated in their bodies, and (perhaps after being released into the environment again for a while) eventually is taken up by some new organism. It is literally true that we are lent the materials for our bodies for a while, these same materials (which may have already been used by many thousands of organisms before us) being later incorporated in other living beings. Understanding these resource cycles is basic to sustainable development and in particular sustainable agriculture [60,61], for it makes crystal clear that trees and vegetables can only grow by incorporating such materials into their bodies; if these elements are not replenished in the earth when crops are harvested, new generations of plants and trees cannot grow for very long.

The way in which energy is channelled through an ecosystem is also vital to its functioning. The basic input source of energy is sunlight, which is absorbed by plants and stored by them in chemical form. This chemical energy is then utilised by animals that eat the plants, and by successive predators that prey on other animals in the food chain [35,58]. The usable energy is eventually reduced to heat, this waste energy being radiated away to space. (If this did not happen, the earth would rapidly warm up and become so hot that life would be impossible [34]; thus the Earth is held in an overall energy balance through its gains and losses of heat.)

In the case both of the materials for life and of the energy that enables it to function, these cycles exist because of the physical conservation laws already discussed. In particular they point up the fact that, since available material and non-renewable energy resources on the surface of the Earth are strictly finite, once these resources have been used (for practical purposes) they will be gone forever. This is the basis of the need for conservation policies, an issue relevant to all of us because of our inter-relatedness.

Interdependence and complexity

The beings in an ecosystem develop intimate relationships resulting in *complex interdependence* . (Flowers feed butterflies which pollinate the plants; foxes depend on eating rabbits who depend on eating carrots and similar vegetables; the air is breathed by animals and recycled by plants and trees; microbes decompose dead bodies, making their materials available again for future use). Interference with one component of an ecosystem can lead to unexpected consequences in quite different components because of these dependencies. While we can try to estimate these effects by use of mathematical models, three sources of uncertainty make this difficult.

Firstly, it is usually difficult to obtain the needed data (how many foxes are there in England? How many whales in the Antarctic ocean? What is the birth-rate for albatrosses in different age groups?) Secondly, the equations in this case represent average statistical behaviour, rather than a precise physical law, so they only give an approximate description of what is likely to happen. (Compare the case of physics where the laws are precise laws enabling us, for example, to predict future planetary positions with great accuracy.) Thirdly, the very complexity of the interactions makes it difficult to solve the resulting equations accurately. In particular, non-linear interactions in the system can lead to chaotic behaviour [64], when very small changes in initial conditions lead to very large changes in the resultant behaviour. This can lead to great uncertainty in the prediction of future growth patterns of interacting populations, because we cannot know the initial data to infinite accuracy.

Similar problems of chaotic behaviour affect weather forecasting, despite the fact that in this case the underlying equations are well founded and can even occur in physical systems such as a forced pendulum. This has caused some excitement as indicating a possible avenue whereby apparently completely deterministic laws (i.e. laws where the resultant of a particular initial state is strictly determined by those initial conditions) can lead to unpredictable results, leading to an 'openness' in physics not previously understood; we then have *deterministic chaos* (despite the equations being deterministic, their results are unpredictable). However it is important to make two points about deterministic chaos. It is an amplifier of existing uncertainty, rather than a fundamental source of uncertainty; this means that it does indeed imply grave prediction problems in systems exhibiting such behaviour, but not that the physics lacks a precise deterministic nature[2]. Furthermore, chaotic behaviour will not occur when feedback control systems are effectively in operation because they are constructed precisely to prevent this happening.

Stability of ecosystems

While feedback mechanisms work to maintain stability in ecosystems, they are not so well ordered and precise as in a single organism, because they result from the interaction of many independent living entities, often in competition with each other for the same limited resources. As a consequence, they are not always successful in maintaining a balance [57].

It is important to mention a contemporary case where previously existing control mechanisms no longer operate effectively, leading to an instability of major significance that will affect us all because of our interdependence. This is the *exponential population growth* that has occurred in the past few hundred years, and is still taking place in many parts of the world. Because of limited resource availability, the resource consumption associated with a rapidly growing population (and consequent inevitable entropy generation) are major threats to the future well-being of humanity and to the ecology of the planet [57,62]. They are the underlying force behind the looming environmental and ecological problems facing us (for example, the hole in the ozone layer that has caused so much concern). These issues simply did not arise when the population was small. They are exacerbated by massive over-consumption by a minority of the world's total population (this being a consequence of major global inequity in resource distribution and use). They can be countered by more caring and conservation-centred resource use policies, leading to a more equitable distribution of wealth. They can also to some extent be compensated for by technological advances, and indeed both measures are needed to improve the situation. However by themselves these steps cannot solve the problems arising from unchecked population growth, which will in the end swamp any gains we can make by such methods. The issue will have to be tackled directly in order to stabilise the earth's population and to prevent continual exacerbation of the regional problems occurring from massive local concentrations [57,62,63].

4.3 Evolution

Apart from understanding in more detail how complex systems work, the major question remaining is how they came into being. In the case of man-made objects, the answer is obvious: they were designed that way, and then manufactured according to that design.

Living plants and animals are immensely more complex than anything men or women have ever designed. How has this been achieved? All the evidence points to a *process of evolution,* whereby single cell animals developed out of carbon-based molecules thousands of millions of years ago, and then evolved to higher and higher levels of complexity, culminating in the existence of the human race [49,65-67]. This interpretation is supported both by the fossil record, and by the present-day observation of evolution taking place in populations of flies, bacteria, viruses, and animals (dog-breeding and horse-breeding is nothing other than evolution of these species, assisted and guided in a specific direction by the breeders). Perhaps most conclusive of all is recently acquired evidence, firstly from developmental biology, where the embryos of humans and of many animals are found to be virtually indistinguishable for the first few weeks of their growth; and secondly from molecular structure and genetics, where the same molecules and molecular mechanisms are known to occur throughout the animal kingdom. In particular the genetic code is virtually universal, controlling the development of fish and birds, of frogs and elephants, of monkeys and men [49,53,65,66].

While the historical occurrence of evolution is beyond reasonable doubt, the nature of mechanisms sufficient completely to account for this remarkable process is still open to debate, despite dogmatic statements on many sides. The biological trend to order proceeds directly counter to the trend to disorder (entropy growth) that is one of the most fundamental characteristics of macroscopic physical systems. At a certain level this is answered by stating that the animals accumulate this order at the expense of their surrounding environment, so that the total system (the animal plus its environment) does indeed obey the entropy law: overall the trend to disorder is obeyed. However this does not explain the mechanisms leading to the increasing order: the source of apparently purposive design.

It is certainly clear that as mutations occur in a population, those animals with a greater ability to survive (because of a greater breeding ability, a greater survival capacity of those born, or both) will tend to dominate the population more and more, while those with a lesser ability to survive will tend to die out. Thus the accepted explanation of evolution is Darwin's mechanism of *natural selection* ('survival of the fittest'), based on:

* *variations produced by random mutation,*

followed by

* *selection based on reproduction and survival rates*, taking place over extremely long stretches of time [65].

This is a powerful feedback mechanism, but with the goal not of production of any particular type of animal, but rather simply of an increase of survival rates; the resulting control process acts as a purposive directional device to increase fitness for survival, with an impressive ability to lead to functional design of living organisms.

The question that remains is whether purely random mutations are a source of variation able to provide a complete explanation of all we see. We do not fully understand the interaction between mutations occurring at the genetic level (which must be their location) and the resulting changes in the animal population. This may be associated with the question of 'punctuated equilibria', where some of the fossil record seems to indicate long periods of very slow change, followed by short bursts of rapid evolution. The situation is complicated by 'piggybacking', where an animal's most obvious physical characteristic was not the feature that led to selection, but came along also as a by-product of some other feature that was the real evolutionary determinant leading to improved survival. In the end the key point is whether evolution of complex structures, for example a wing, can *always* be achieved by a route involving many thousands of small purely random mutations, in which *every step* of the way leads to an improved survival rate. This is despite the fact that until it is developed enough actually to enable it to fly, a developing wing is probably a hindrance rather than a help to the animal in its fight for survival. Dawkins [65] strongly argues that natural selection based on purely random mutations suffices; the sceptic desires something rather more in the way of solid proof. The alternative hypothesis is that not all steps in the process are completely random; some kind of direction is given to mutations that take place, so the process is not purely based on chance. This could occur, for example, if the nature of physical and chemical potentials preferred particular biological structures to others (which certainly is the case when elementary biological molecules form from some primeval soup of chemicals; the question is how far this effect extends).

However these are controversial issues; for example the apparent changes of rate of evolution might simply be due to selection effects (which determine what fossils are actually available to us for examination, as remnants from the many millions of creatures that have lived in the past). Certainly conclusive discussion of such issues is hindered by gaps in the fossil record, forcing us to guess what the situation was from a fraction of the evidence that might in principle be available to us. In any case, this line of questioning does not doubt the existence of evolution, nor the functioning of the mechanism of natural selection. But it queries whether all the mutations that take place are completely random, or whether they are in some sense directional towards their final product (for whatever reason).

Perhaps most difficult of all is the question of how life first arose. There are a series of sticking points in the evolution of living systems, each of which is difficult to overcome; perhaps the most difficult of all is the origin of the first functional cell, complete with sufficient DNA and RNA to replicate. Ingenious theories have been proposed, such as Manfred Eigen's that natural selection can operate in particular inorganic systems in such a way as to lead to living systems, or the idea that silicates could provide a 'scaffold' on which complex organic molecules could then form. The problem again is to introduce some kind of direction into the process so that it will proceed with more certainty than purely random variations. However none of these ideas has achieved universal acceptance; essential steps remain unresolved [65,68].

Given that evolution has taken place, it is interesting [65] to think of the space of all possible animals allowed by the laws of physics and chemistry (that is, all the animals that could possibly exist and function), and then the subspace of that space that has been actualised through the historical process of evolution. That subspace is very large (cf. the enormous variety of animals around us [49]); however the whole space of possibilities may be enormously greater. In any case, one of the most spectacular achievements has been the evolution of consciousness and language, and subsequently of social and economic systems, the whole nature of evolution changing with the advent of consciousness. Intimately tied in with this is the fundamental question of the evolution of ethics and morality, and of a sense of aesthetics. These issues will be developed in Chapter 8.

4.4 Complex Organisation

Complex functionality is made possible by a hierarchical structuring of complex systems (motorcars, aircraft, computers, etc.). In particular, this is the basis of order in living systems [49], as summarised in Table 3.

Organic Molecules
Cells
Tissues
Limbs and physiological systems
Individual organisms and animals
Animal populations
Ecosystems
The biosphere

Table 3: Hierarchical nature of living systems

Considering how such systems function, we conclude that:

> The particles and forces considered in the previous chapter provide the basis for complex organisation, and in particular for the functioning of living organisms, through their ability to form hierarchical structures that can store and utilise information in hierarchical feedback control systems.

In particular, one of the most amazing information stores is the molecule DNA, containing digitally coded instructions for constructing living organisms. This information contains within itself a historical record of the evolutionary process that has led to our existence; it is our genetic heritage resulting from that process.

CHAPTER FIVE

THE PHYSICAL UNIVERSE

This chapter provides an overview of our current understanding of the nature of the physical Universe. In turn we consider the large-scale distribution of matter in the Universe, its present expansion and past evolution (the history of the 'Hot Big Bang') and its possible future. This is the larger environment in which we exist.

The biosphere exists and flourishes on the surface of the Earth and in its immediate environment (the rivers, lakes, and seas, and the surrounding atmosphere), depending on them for its existence, so this is the immediate context in which our own existence is based. We now focus on the broader environment that makes all of this possible: the Solar System, Galaxy, and surrounding Universe, whose nature we explore by using telescopes of various kinds [69,70]. This astronomical environment has been the subject of speculation and study for thousands of years [71,72]; the past three decades have led to unprecedented new understanding of its nature [73,74].

5.1 The geography of matter

The planet Earth[1] on which we live is a sphere about 12,500 km diameter, with a solid crust made of rock, dust and soil at a temperature[2] of about 300 K, overlying a molten core made largely of iron at a temperature of about 3000 K, and surrounded by a thin layer of air (about 100 km thick) composed mainly of nitrogen, oxygen, and carbon dioxide [75]. It is held together by gravity, which we feel as a force pulling us towards the centre of the earth. Two-thirds of its surface is covered by water (the seas), resulting in much water vapour circulating through the air and falling back to the surface as rain or snow. Its surface is in a state of continual change and turmoil [67]; much of this is only appreciable on geological time-scales, much longer than a human lifetime, but it is also manifest to us in volcanoes and earthquakes [76].

The local Region

Earth is surrounded by empty space. It moves in a nearly circular orbit around the Sun, 150 million km away, being held in this orbit by the Sun's gravitational attraction, and taking a year to complete each cycle (Copernicus' great discovery [71]). The center of the Earth is heated by decay of radioactive materials, but the energy source for the biosphere (and so in particular for our lives) is heat radiated to the Earth by the Sun, captured and turned into chemical energy by plants and trees.

The Sun is very much larger and more massive than the Earth [69]. It is a sphere of very hot gas (mainly hydrogen and helium), with a temperature of 15 million K at the centre and 5,500 K at the surface. The energy source for the Sun is the nuclear fusion of hydrogen nuclei to form helium near its centre (it is a giant free-floating fusion reactor, held together by gravity), the heat generated being radiated away to space from its surface [77,78]. The Earth is one of nine planets circling the Sun, the whole collection (together with asteroids and comets that also orbit round the Sun) forming the Solar System. The Sun is the central object in the Solar System, containing most of its mass and providing virtually all the energy pervading the region. Despite appearing so different from them, the Sun is a typical star [77], looking much larger and brighter than all the other stars simply because it is nearer to us than they are (the nearest star - after the Sun - is about 4 light-years away, a light-year being about 90 million billion km)[3].

All the bright stars we see around us belong to the *Galaxy*, a rotating disk of stars and dust about 60,000 light years across[4] with spiral arms surrounding a central region; the disk is visible to us as the *Milky Way* (the band of stars clearly visible in the night sky when one is well away from city lights) [69]. Within and around the Galactic disk there are thousands of star clusters as well as huge clouds of dust and gas. Our Solar System is situated in the Galactic disk, towards its outer edge, and circles around the centre of the Galaxy in an orbit that takes about 250 million years to complete. There are about 100 billion (100,000,000,000) stars in the Galaxy, each one something like the Sun; many of these stars may be surrounded by planetary systems like our own Solar System. It is very difficult to detect such planets from Earth, because they do not shine like a star does; their surfaces are only lit by reflection of light from the central star, which will far outshine them. However if there is life elsewhere in the Galaxy, it will have developed on the planets in such planetary systems (the stars themselves being far too hot to support life).

The larger domain

Faint wisps of light that we can detect between the stars in the night sky turn out to be other galaxies, that are enormously far away (the nearest is about a million light years away). One of the great triumphs of astronomy has been to determine the distances of these objects, thereby showing that they are in fact huge star systems as large as our own Galaxy [69,72]. We are able to detect about 100 billion such galaxies around us, each containing about as many stars as our own but appearing extremely faint and small because of their distance from us; we can also detect many other very distant objects such as X-ray sources, radio sources, and quasi-stellar objects [70], many of which may be galaxies at various stages of their evolution. They appear to extend on without end, together forming *the Universe;* that is the entirety of that unique interacting distribution of matter and radiation that is the basis of our physical reality [79-81]. As discussed later, the observable region of the Universe is almost certainly a small part of the entire Universe. *Cosmology* is the subject that studies the distribution and history of matter on these vast scales.

In the region round us, galaxies seem to be organised in clusters which in turn form huge walls surrounding vast voids [80]. On the very largest scales (say greater than 300 million light years) galaxies appear to be isotropic about us (i.e. the distribution of matter looks the same in all directions we observe). Indeed all the broad features of the Universe appear to be the same everywhere; we cannot for example point to any particular region as being its centre. Consequently we believe matter and radiation are distributed uniformly in space, i.e. the Universe is *homogeneous* on the largest scales [79,80], so that conditions are the same in every place [5.]

In summary, on astronomical scales, the matter we can see is hierarchically organised into the structures listed in Table 4.

Planets, including the Earth
Stars and their planetary systems
Galaxies
Clusters of galaxies
Large scale structures (walls, voids)
The observable Universe

Table 4: Hierarchical nature of astronomical structure

It is difficult not to get overawed and perhaps even dismayed on realising the prodigious number of astronomical objects, their size, and their distance. Humanity seems truly insignificant in the face of these vast scales. Certainly the Solar System is physically tiny compared with the size of the Local Cluster of galaxies, which in turn is just a tiny part of the observable region of the Universe. However physical size is not the only criterion of importance or of causal significance (this is clearly true on Earth; it is just as true in the Universe as a whole).

5.2 The history of the Universe

The major issue in physical cosmology is, where has all this matter come from, and how has it come to be organised in this particular way? [82,83]. There have been two major viewpoints: that the Universe is unchanging in time, or that it is evolving (i.e. major changes take place in its physical state as time unfolds) [84,85].

The view put forward by Einstein in 1917 when he formulated the first quantitative universe model was that the Universe is static, that is, is completely unchanging in time (when viewed on the largest scales). This was indeed taken for granted by most of the cosmologists of the period. However it was difficult for them to explain the spectra of light received from distant galaxies, which show systematic *redshifts* (i.e. a displacement towards the red of spectral lines, the displacement being proportional to wavelength), with the amount of redshift increasing proportionally to the distance of the galaxy. Such redshifts do not occur in the Einstein static universe[6]. Furthermore it was shown that the Einstein static universe model is unstable, so one could not expect the Universe to remain in this static state.

Consequently the viewpoint changed [72,85]; the spectral redshifts are now interpreted as being caused by motion of galaxies away from each other (redshifts arising naturally from such a recession). Thus the idea of the *Expanding Universe* became established. The expansion is universal in that every galaxy recedes equally from every other one (there is no centre to the expansion, because the Universe is spatially homogeneous).

At first glance this seems to imply we must abandon the idea that the Universe could be unchanging. However this is not necessarily so, because of one rather exceptional kind of expansion that can occur if the Einstein Field Equations are modified appropriately. This is when the Universe is always expanding, but the rate of expansion and the density of matter are always the same; clearly this requires a continuous creation of matter to keep the density constant while the Universe continually expands. Such a universe is called a *Steady State Universe*. For many people this is an attractive possibility, because an unchanging Universe is thought to be 'more perfect' than a changing one [79]. However this model cannot explain evidence obtained from optical and radio telescopes that there was a higher density of radio sources and quasars in the past than at present (the densities would have to be unchanging, if the Universe were in a steady state, for then conditions would be the same everywhere); and it does not give a natural explanation of the cosmic background radiation (discussed below). It has therefore been abandoned by almost all cosmologists.

The Evolving Universe

The present consensus, then, is that the *Universe is evolving,* having had a higher density of matter in the past (when the galaxies were closer together) than at present [79,86-91]. This is one of the major discoveries of cosmology. It has profound consequences: if we apply Einstein's gravitational equations to an evolving universe model, assuming that the behaviour of matter is normal, we find that the Universe must have originated in a singular state where the density and temperature were infinite (follow the motion of the galaxies backwards in time; they get closer and closer together until the matter of which they are made reaches an infinite density[7] . That is, the implication is not only that the Universe is evolving, but that *it had a beginning a finite time ago.* Furthermore, we can determine when this happened: measurement of the rate at which the recession velocity increases with distance (*Hubble's constant*) gives an estimate of about 10 billion years for the age of the Universe.

The Hot Big Bang

Because the matter gets hotter and hotter as we go back in time towards this initial state, we can talk of the *Hot Big Bang,* and use standard physical laws to examine the physical processes going on in the high density and temperature mixture of matter and radiation in the early Universe [86-89]. An important feature is that when the temperature dropped lower than about 3000 K as the Universe expanded, nuclei and electrons combined to form atoms; but at earlier times when the temperature was higher, atoms could not exist, as the radiation then had so much energy that it disrupted any atoms that tried to form into their constituent parts (nuclei and electrons). Thus at early times matter was *ionised,* i.e. it consisted of electrons moving independently of atomic nuclei. Under these conditions, the free electrons interact strongly with radiation. Consequently matter and radiation were tightly coupled together by scattering processes, and the gas in the Universe was opaque to radiation (rather like the interior of the Sun). When the temperature became lower so that atoms formed from the nuclei and electrons, this scattering ceased and the Universe became transparent (today we are able to see galaxies at enormous distances from us). The time when this transition took place is known as the *time of decoupling* (it was the time when matter and radiation ceased to be tightly coupled to each other).

Radiation was emitted by matter at the time of decoupling, thereafter travelling freely to us through the intervening space. When it was emitted, it had the form of Black Body radiation at 3000K. As it travelled towards us, the Universe expanded by a factor of 1000; consequently by the time it reaches us, the radiation has cooled to 3 K (that is, 3 degrees above absolute zero, with a spectrum peaking in the microwave region), and so is extremely hard to detect. *This Cosmic Microwave Background Radiation* was detected in 1965, and its spectrum has since been intensively investigated, its black body nature being confirmed to high accuracy; this is now taken as solid proof that the Universe has indeed expanded from a hot big bang. The radiation is often referred to as Relic Radiation from the Big Bang. An important feature is its high degree of *isotropy:* for its temperature is the same in all directions about us, to better than one part in 10,000. This is the major reason we believe the Universe is uniform and isotropic (any inhomogeneities or anisotropies in the matter distribution lead to anisotropies in this radiation, as recently discovered at a very low level by the extremely sensitive detectors of the COBE satellite).

The history of matter

Before decoupling, the temperature of the Universe exceeded any temperature that can ever be attained on Earth or even in the centre of the Sun; as it dropped towards 3 K, successive physical reactions took place that determined the nature of the matter we see around us today. An important time was the era of *nucleosynthesis,* the time when the light elements were formed. Above about a billion degrees K, nuclei could not exist because the radiation was so energetic that as fast as they formed, they were disrupted into their constituent parts (protons and neutrons). However below this temperature, once these particles had collided with each other with sufficient energy for nuclear reactions to take place, the resultant nuclei remained intact (the radiation being less energetic and hence unable to disrupt them). Thus the nuclei of the light elements (deuterium and tritium [8], helium, lithium) were created by neutron capture. This process ceased when the temperature dropped below about 100 million degrees K (the nuclear reaction threshold). In this way, the proportions of these light elements at the time of decoupling were determined; they have remained virtually unchanged since. The rate of reaction was extremely high; all this took place within the first three minutes of the expansion of the Universe [86-88]. One of the major triumphs of Big Bang theory is that the predicted abundances of these elements (25% Helium by weight, 75% Hydrogen, the others being less than 1%) agrees very closely with the observed abundances. Thus the standard model explains the origin of the light elements in terms of known nuclear reactions taking place in the early Universe.

In a similar way [89-91], physical processes in the very early Universe (before nucleosynthesis) can be invoked to explain the ratio of matter to anti-matter and the ratio of matter to radiation in the present-day Universe. However other quantities (such as electric charge) are believed to have been conserved even in the extreme conditions of the early Universe, so their present values result from given initial conditions at the origin of the Universe, rather than from physical processes taking place as it evolved. In the case of electric charge, the total conserved quantity appears to be zero: there are equal numbers of positively charged protons and negatively charged electrons, so that at the time of decoupling there were just enough electrons to combine with the nuclei and form uncharged atoms (it seems there is no net electrical charge on astronomical bodies such as our galaxy; were this not true, electromagnetic forces would dominate cosmology, rather than gravity).

After decoupling, matter formed large scale structures (as discussed in the next section) which eventually led to the formation of the first generation of stars. However at that time planets could not form for a very important reason: there were no heavy elements present in the Universe. The material out of which this first generation of stars was made consisted mainly of hydrogen and helium, with a trace of deuterium, tritium and lithium, for this was the result of the process of nucleosynthesis in the early Universe; planets cannot be formed out of these elements alone (nor can complex beings such as humans). The first stars aggregated matter together by gravitational attraction, the matter heating up as it became more and more concentrated, until its temperature exceeded the thermonuclear ignition point and nuclear reactions started burning the hydrogen to form helium (like in the centre of our Sun). Eventually more complex nuclear reactions started in concentric spheres around the centre, leading to a build up of heavy elements (carbon, nitrogen, oxygen for example) [77,78]. These could not form in the early Universe because the whole process there took place so swiftly, but they can form in stars because there is a long time available (millions of years).

Massive stars burn relatively rapidly, and eventually run out of nuclear fuel. At this time a dramatic event happens: the star becomes unstable, and its core rapidly collapses because of gravitational attraction. The consequent rise in temperature blows the star apart in a giant explosion, during which new reactions take place that generate elements heavier than iron; this explosion is seen by us as a *Supernova* ('New Star') suddenly blazing in the sky, where previously there was just an ordinary star (or perhaps the star was so faint we saw nothing there at all). This is fundamentally important to us, because such explosions blow into space the heavy elements that had been accumulating in the star's interior, forming vast filaments of dust (such as those we see in the Crab Nebula) around the remnant of the star. It is this material that can later be accumulated, during the formation of second generation stars, to form planetary systems around those stars. Thus the elements of which we are made (the carbon, nitrogen oxygen and iron nuclei) were created in the extreme heat of stellar interiors, and made available for our use by supernova explosions. Without these explosions, we could not exist.

Overall, we obtain a dramatic and compelling picture of the rapid evolution of the primeval fireball that comprised the early Universe, cooling down from extreme temperatures as the matter and radiation expanded. The primordially existing gas, initially composed of an equilibrium mixture of particles and radiation, underwent a series of irreversible processes: baryons (essentially, protons and neutrons, providing the materials to make nuclei) were synthesized out of the elementary particles that dominated the earliest epochs; light nuclei were synthesized out of the baryons; decoupling of matter and radiation took place; first stars formed (powered by gravitational attraction). Heavy elements were formed in first generation stars. In this way we can understand the origin of the matter around us, and the chemical elements on Earth and in the Sun.

5.3 The origin of structure

While the basic Big Bang picture is clear and is strongly supported by evidence, the way astronomical structures have arisen is not so clear.

At the very largest scales, there is still a basic unresolved dichotomy: we do not know if the Universe started in a very smooth state and then developed inhomogeneity, or started off very inhomogeneously and then became very smooth. Either way, our problem is to understand how the Universe at present appears so smooth on the very largest scales (particularly as evidenced by the isotropy of the microwave background radiation), and yet has complex structuring on all smaller scales, as outlined in the previous section.

The Inflationary Universe

The original viewpoint was that the Universe started off very smoothly, that is the density and temperature were almost uniform everywhere. Then small perturbations grew by gravitational attraction to form the large-scale structures we see. However the origin and nature of the required perturbations was a mystery, as was the reason the Universe appeared so smooth on the very largest scales. Consequently the *Chaotic Cosmology* idea was advanced, assuming that the Universe started off very inhomogeneously in the large but then physical processes caused it to develop into a very smooth state overall. This idea received tremendous impetus when Alan Guth, Andrei Linde and others pointed out that if a particular form of matter (a *scalar field,* associated with particles without spin) dominated the expansion of the early Universe, there could be an enormous expansion ('inflation') in a very small period before nucleosynthesis - an expansion much more rapid and of a much greater magnitude than would occur in an ordinary Big Bang model [20,88-91].

The result of this expansion is a dramatic smoothing of the largest scale inhomogeneities - like a balloon being blown up, becoming smoother and flatter as it gets larger and larger. So this *inflationary Universe* idea explains the overall smoothness of the Universe as the result of physical processes. This has caused much excitement and a dramatic surge of activity while the consequences of the idea have been worked out. It has in particular led to an interesting proposal: the seeds that form galaxies later on could come from quantum fluctuations in the very early Universe, blown up to enormous size by the expansion that takes place during inflation (superimposing a small roughness on the basic smoothness). This enables us in principle to link the present distribution of galaxies with the properties of elementary particles in the very early Universe: a truly remarkable claim.

While the idea has many supporters, it is not without problems, not least being that the particular field which is supposed to underlie inflation has not been identified. Furthermore Roger Penrose [34] argues strongly that this model cannot be correct because of the Arrow of Time problem (discussed below), which he argues demands a smooth start to the Universe. Initial smoothness is also an intrinsic part of the Hartle-Hawking Quantum Cosmology programme (briefly outlined below, and discussed in Hawking's famous book *A Brief History of Time* [92]). Thus this basic issue is at present unresolved.

The formation of local structure

Assuming that the Universe has reached a state where it is very smooth in the large, the next question is how galaxies and larger scale structures arose from initial seeds of inhomogeneity. Here again we have many proposals but no certainty, and a basic dichotomy between two different approaches [20,83,90,91]. On the one hand it is possible that these structures were created *top-down,* with the largest structures (walls and voids) forming first, smaller structures (clusters of galaxies and then galaxies) then coalescing out. On the other hand there could have been a *bottom-up* process, whereby globular clusters of stars formed first, and then later aggregated together to form the larger structures (galaxies and then clusters of galaxies). The first is more likely if the Universe is presently dominated by Hot Dark Matter, and the latter if it is dominated by Cold Dark Matter (discussed below).

Various models proposed are partly satisfactory, but none is compelling; indeed we are not sure if the basic mechanism underlying the start of formation of structures on these scales is gravity alone, or if other processes were significant. For example, vast explosions in the very early Universe, like Supernovae, could possibly sweep up matter around them and so explain the voids; or the creation of vast sheets of galaxies could perhaps be explained by exotic structures known as *cosmic strings* (energy concentrated on extremely thin string-like regions) - if they exist. However once structure has started to form by whatever method, gravity will then amplify the inhomogeneities, in particular forming galaxies.

Within galaxies, first generation stars form by gravitational attraction; some of them die as supernovae, and then second generation stars form from the debris, sometimes with planetary systems (as discussed above). We are uncertain about the nature of the creation of the Solar System. There are two competing theories. One is that the Solar System had a hot origin, the planets being formed from hot gas which coagulated into a disk rotating around the Sun, which has been cooling down ever since, forming the solid planets as molten material cooled down. The other is that the planets had a cold origin, forming from dust accumulating and eventually heating up to give the molten core at the centre of the Earth. Whatever its mode of origin, the Earth was formed about 5 billion years ago, at about the same time as the Sun (the oldest rocks we have found, have been measured to be this old).

Like all other objects in the Universe, the Earth is not static but is rather in a dynamic state of change [67,75,76]. Volcanic eruptions were once far more frequent, when the features on the surface of the Earth were being formed. Mountain chains are not unchanging but rather are made by massive uplifting movements of rock, and thereafter are subject to a steady process of erosion by wind, rain, and ice that shapes their form. Even the continents themselves are not static, but are subject to *continental drift* as massive underlying rock plates slowly move apart (for example, Africa and South America were once adjacent to each other, but are now far apart). The formation of the seas was a necessary precursor to the evolution of life, and as life developed it changed the nature of the atmosphere (at early times the atmosphere consisted mainly of gases like carbon dioxide and sulphur dioxide, and would have been poisonous to us; now it is mainly oxygen and nitrogen). Organic molecules would have been available in abundance before life began (many have been detected in interstellar gas clouds in space).

It is this complex chain of events that prepared the seas and the surface of the Earth as the environment within which life could develop [49,65,68], starting as single cells and then evolving through small multi-cellular animals to fish, to mammals, (with some dead-ends occurring, like the dinosaurs that once used to dominate the Earth but then died out completely, in a mass extinction whose cause is still open to debate). In this way the highest levels of complexity and order known to us were created, leading eventually to humanity and the evolution of consciousness. However it must be emphasized that (as mentioned in the last chapter) various steps along the way, in particular the creation of the first living cell, are not understood [68].

The probability of life

One of the most interesting questions we can ask is whether or not there is life on other planets in our Galaxy, or elsewhere in the visible Universe. This has been the subject of ongoing debate, with some protagonists believing it is very probable, and that there are likely to be millions of planets in the Galaxy with life on them; while others maintain we are the only self-conscious beings in the entire visible Universe [93-95].

The problem in making the estimates is the many uncertainties not only in the probability of formation of a planet like Earth, but of the evolution of life (first single celled forms, and then more complex forms,) once conditions on the surface of the planet had settled down enough for living systems to have a chance to survive. A view held strongly by some is that there are so many improbable steps in the evolution of complex life forms, that life cannot exist anywhere else in the Universe except on the Earth. In biological terms this places the human race at the centre of the Universe, by stating that the Earth is the one unique planet on which intelligent life has evolved (despite the fact there are something like ten thousand billion billion stars in the observable region of the Universe - and could be an infinite number in the unobserved regions).

In my view we should resist this return to a pre-Copernican view, with ourselves the only conscious beings in the entire Universe and therefore the highest state of organisation that exists, unless it is absolutely necessitated by proper estimates of all the probabilities involved; however these simply are not known at present. Until we are forced to conclude otherwise, I would suggest that the presumption must be that we are not unique; the laws of physics and chemistry that created life here would also have done so elsewhere, given this vast array of stars (a reasonable proportion of which must have planetary systems where the same processes that operated to create life here would be at work).

A final comment on this topic: much has been made of the argument that there cannot be intelligent life elsewhere in the Galaxy that has reached a technological stage, for if there were they would already have visited us [95]. In my view this is an interesting argument but should not be confused with a scientific proof (as some have tried to present it). It has at least two major flaws: firstly, the presumption that we know what the intentions of such other intelligent beings would be. Other beings with the capability to do so might choose to spend time and effort on exploring the Galaxy, but then again they might not. The attempt to claim that they inevitably would do so is sociologically and psychologically naive[9]. Secondly, it would take a minimum of 200,000 years to explore the Galaxy (and to do so thoroughly could take enormously longer); so even as we are proclaiming that they do not exist, they may be heading towards us to investigate the radio signals we have been broadcasting for the past sixty years. Even disregarding the sociological issues, the 'proof' cannot be rigorous. If there is any group of beings in the Galaxy that, having reached an advanced technological stage, decides to explore the Galaxy, then there must be a first group to do so; they are the first beings to set out to explore the galaxy. The argument is clearly fallacious when applied to that first group.

The structural options

Overall, we see in this discussion much more uncertainty than in the previous two chapters. The reason is straightforward: we are here dealing with historical science, rather than analytic or synthetic science (cf. Chapter 2). Thus there are a series of unresolved major dichotomies (Table 5); in each case we have many studies of possible modes of evolution, and some indications of which is correct, but no final answers.

Structural Options

Structure	Option 1	Option 2
The Universe as a whole	From smooth to structured	From Chaotic to smooth
Large scale structures	Top-down (Hot Dark Matter)	Bottom-up (Cold Dark Matter)
Solar system	Hot origin	Cold origin
Intelligent life	Very Rare	Quite Common

Table 5: Alternative views on the evolution of structure in the Universe

However despite these uncertainties, we have attained a good degree of certainty about many aspects of what has happened. One feature is well recognised: it is gravity that primarily powers the generation of astronomical structures [34]. The objects about us are essentially the results of non-equilibrium events as the universe expanded from a relatively uniform state (at the time of decoupling) and matter attempted to cluster itself more and more because of gravitational attraction. with various forces successively slowing the process down ([34], Dyson in [35]). Thus while the general tendency implied by the Second Law of Thermodynamics is towards disorder, gravitation creates structures at the astronomical scale, and evolution does so in the biological world.

5.4 Final states

Having attained a partial vision of origins, the issue of endings or final states is also of interest. Predicting the future is of course even more hazardous than charting the past. Nevertheless there are some things that we can say about it with a high degree of certainty, and some cases where we can clearly lay out a number of major options with quite different implications.

Stars and Black Holes

One thing we can say with absolute certainty is that the Sun is going to stop shining. The reason is that it has a finite stock of nuclear fuel available, which will eventually run out. The more massive a star is, the more quickly it evolves to its end-state; our own sun is not very massive (as stars go), and will probably survive for another thousand million years or so. At that point, as is the case with any star, there are a number of options available [77,78]. Depending primarily on its mass, a star can

(1) just fizzle out, ending up something like the planet Jupiter;

(2) have a dramatic collapse (perhaps involving an explosion which we see as a Supernova), and end up at a very condensed stable state, becoming either a *White Dwarf* (a star supported by the degeneracy pressure that results from the Pauli exclusion principle) or a *neutron star* (where the whole stellar interior is like one giant nucleus); or

(3) collapse to a black hole.

A *Black Hole* occurs when space-time becomes so highly curved, due to the high gravitational fields involved, that it traps everything in its vicinity and drags them into itself, arbitrarily large tidal forces then destroying anything that falls in [31,34,79]. As light tries to escape from the interior of a black hole it is pulled back by the massive gravitational field, and falls in; consequently we cannot see to the interior (light cannot convey information to us from the interior, because it cannot reach us). Classically considered, at the centre of the black hole there is a breakdown of the structure of space-time itself (*a space-time singularity*); however this is hidden from us by the *event horizon* (the surface separating events from which light can escape the black hole, and those from which it cannot). A black hole eventually radiates its mass away as black-body radiation, because of a remarkable quantum mechanical process discovered by Stephen Hawking, the rate of radiation depending on its mass.

We have seen many white dwarfs and neutron stars, the latter being detected as *pulsars* like that at the centre of the Crab Nebula, emitting periodic radio signals with incredible precision. Precisely because of their nature, black holes are difficult to detect; however infalling gas will get heated up until it emits radiation which can be detected, and we can also hope to detect them by their gravitational effects. On theoretical grounds we believe that there are some hundreds of thousands of black holes in our galaxy, and some reasonably good black hole candidates have been identified, for example the X-ray source Cygnus X-1; there may even be a massive black hole at the centre of the Galaxy. Evidence that these identifications are correct is strong, but not overwhelming.

If a black hole came near the Earth, the effects could be devastating; however there is no reason to believe there is one near enough to us to pose any threat. The asteroids in the Solar System are far more problematic (the craters on the Moon are evidence of massive bodies colliding with it in the past), but we have not detected one on a collision course with the Earth. The real problem for the Earth (in the long term) is that when the evolution of the Sun is very advanced, it will probably expand to become a Red Giant, enveloping us in its hot atmosphere. If this gas does not extend far enough to vaporise the Earth's atmosphere, seas, and surface, then the next phase will do so, for the Sun will probably then blow off its outer layers in a massive explosion (forming a Planetary Nebula), while the core collapses to form a white dwarf star. The eventual cooling of the remnant of the Sun will cause any surviving planets to become bitterly cold, their heat source having died out; they will eventually cool to a few degrees centigrade (the temperature of the background radiation). This scenario is fairly certain: the time when it will occur is not well determined, but is certainly a very long way off.

The Universe

The final state of the Universe itself has also been the subject of speculation and investigation [96,97]. There are essentially two options for the present phase of expansion. Either

* the Universe will continue to expand forever, or

* it will reach a maximum size, and then recollapse to a hot dense state in the future.

In the first case, as the Universe continues its eternal expansion, all the stars and galaxies eventually cool down as all the nuclear fuel is used up, the stars running their life cycles and ending up as white dwarfs, neutrons stars, or black holes. These themselves are probably all unstable too in the long term with not only the black holes radiating away by the Hawking process, but all the other stellar remnants decaying (because baryons are unstable on a very long timescale). This is the first form of *heat death* envisaged for the Universe[10]: the Second Law of Thermodynamics triumphs as everything eventually cools down and radiates away, leaving nothing but the emptiness of interstellar space, with even the last glimmers of light from cooling neutron stars eventually fading out to leave an eternal blackness.

In the second case, by contrast, as the Universe collapses, everything heats up indefinitely and we have in effect a re-run of the Big Bang, but run backwards in time. The matter and radiation in the Universe get hotter and hotter, all the structures that have been built up get destroyed by the increasingly hot radiation (remaining stars explode because they can no longer radiate away their energy to the sky, which is hotter than their surfaces; molecules, atoms, even nuclei are decomposed into their constituent particles by the ceaseless bombardment of radiation). Entropy triumphs as the Universe races to increasing disorder, and accelerates towards a space-time singularity in the future, or at least a region where classical physics breaks down completely because of the extreme conditions prevailing. This is a heat death of the second kind. It could not occur before the Universe is at least ten times as old as the present, and could conceivably take much longer than that to occur; however once started, the final phase (like the Big Bang) will be extremely rapid.

Which of these will actually happen? This is one of the great unsolved questions in cosmology: we do not know. The issue hinges on how much matter there is in the Universe [90,91]. If there is a lot of matter present (in terms of the *density parameter* Ω[11] measuring the density relative to the square of the Hubble constant, this is when Ω is greater than 1), then the gravitational attraction of matter will eventually win over the kinetic energy of the expansion: the Universe will halt its expansion and recollapse. If Ω is less than 1, the kinetic energy will win and the Universe will expand forever. When Ω equals 1, there is a *critical density* of matter present; in this borderline case, the matter just succeeds in expanding forever. Thus the question is, what is the value of Ω?

Now the visible matter in the Universe corresponds to a value for Ω of about 0.04, and dark matter whose existence can be deduced from its observed gravitational effects amounts perhaps to Ω of at most 0.2. Taken at their face value, this implies that we live in a low density universe that will expand forever. However many people maintain Ω is very close to 1. Why should anyone suppose that Ω takes this value, implying that more than 96% of the matter in the Universe is dark matter, unobserved by us? It is easy to conceive of matter that is hard to detect (for example, small rocks distributed through space), but why should it exist? The major reasons for taking the possibility seriously are theoretically based.

The Dark Matter Enigma

Firstly, it is usually supposed that the inflationary model (discussed above) necessarily implies that the Universe expanded so much in early times as to drive the density parameter arbitrarily close to 1, so that it still remains very close to that critical value today (the theory does not say if it should be ever so slightly over or under this value). This then implies the existence of huge amounts of dark matter, dominating the dynamics of the Universe.

Originally this high density was thought to be prohibited because it would ruin the very good agreement of nucleosynthesis calculations with observations. However it has been pointed out that this is not necessarily so if the dark matter is *non-baryonic*, that is it is made of exotic particles (massive neutrinos, magnetic monopoles, axions, and so on) whose existence is predicted by elementary particle theory, rather than the protons and neutrons that are the substance of ordinary matter. A large number of proposals have been made for such dark matter candidates [90,91]. An important controversy is whether they are massive so that they cooled down early, thereafter forming *cold dark matter*, moving slowly at the time of galaxy formation (and resulting in a bottom up process of large scale structure formation), or have a low mass and cooled slowly, therefore for a long time forming *hot dark matter*, moving very fast at the time of galaxy formation (and resulting in a top-down galaxy formation scenario). These possibilities are presently being hotly investigated and debated [73,74,90]; while there is no definitely agreed evidence that such large amounts of dark matter exist[12], it is certainly possible that they do.

Secondly, an important feature of a high-density universe with Ω greater than 1, is that it necessarily has closed (finite) spatial sections [31]. If we set off in any direction in space and just keep going, we would end up back where we started, rather as happens on the surface of the Earth when we steer an undeviating course (although the surface of the Earth is a two-dimensional surface; in these cosmologies, the same effect occurs in the curved 3-dimensional space sections of space-time that are surfaces of constant time for all the fundamental observers [31]). There is therefore necessarily a finite amount of matter in such a universe, and a finite number of galaxies exist. By contrast, in universes with Ω less than 1, unless the connectivity of the space-sections is unusual (this case will be discussed in the following chapter) the space sections are infinite. When one considers this carefully, it is a rather difficult feature to come to terms with: there are an infinite number of galaxies in such a Universe, an infinite amount of matter is created at the Big Bang, there are an infinite number of living beings in it. Albert Einstein and John Wheeler have pointed out other advantages of closed space sections, specifically that this does away with the problem of setting boundary conditions for physical fields at infinity (for there is no spatial infinity in such Universes). Wheeler claims that these advantages are so overwhelming that such Universes are necessarily to be preferred over all the others, indeed that the Universe must be this way.

Neither argument is water-tight. Inflation may not have taken place, and if it did, one can query whether if this necessarily predicts a high density of matter. Strong as the boundary condition arguments for a closed space section are, they have left many physicists unconvinced; furthermore there are some low density Universes that also have closed space sections. The issue must be regarded as unresolved: it remains one of the major challenges to cosmologists to determine if the Universe will recollapse in the future, or not. If we can determine the density of matter in the Universe accurately enough, it will give us the answer.

Phoenix Universes?

The major further question is whether the process of expansion only happens once in the life of the Universe, or occurs repeatedly.

The first option is the standard model, where the entire evolution of the Universe is a once-off affair, with all the objects we see, and indeed the Universe itself, being transient objects that will burn out like dead fireworks after a firework display. In this case everything that ever happens occurs during one expansion phase of the Universe (possibly followed by one collapse phase, as discussed above).

The alternative is that many such phases have occurred in the past, and many more will occur in the future; that is, the Universe is a *Phoenix Universe,* new expansion phases repeatedly arising from the ashes of the old. This could conceivably occur in a spatially homogeneous way, the entire Universe undergoing successive phases of expansion and collapse (demanding a high-density with Ω greater than 1 to cause the collapse), followed by re-expansion; or it could occur inhomogeneously, with small parts of the Universe collapsing and then rebounding to form the seeds of vast new expansion phases. Such cyclic universe models combine the advantages of predicting hot big bang evolutionary phases, consistent with our present observations, but also potentially having an eternal character by allowing an overall repetitive pattern of behaviour that might possibly go on for ever.

While the idea is an old one, repeatedly proposed, actual mechanisms that might allow this behaviour have not yet been elucidated in detail in a fully satisfactory way. It is possible that quantum theory will come to the rescue and show that some form of continuation is possible in either case. If we consider the recollapse of the Universe as a whole in the future, it is possible that eventually (when conditions were very extreme, for example much hotter than the temperature of nucleosynthesis) quantum fields would eventually dominate and turn the collapse around to create a new expansion from the ashes of the old with the entire Universe re-expanding. Whether the Universe as a whole recollapses or not, it is possible that because of such quantum processes the local collapse of matter to form a black hole will result, not in a singularity, but in a re-expansion into a new Universe region to form the seed for a new big bang phase. If so, this process would not be visible from the old universe region (it would be hidden behind the event horizon), so that as the old expansion phase decayed away, the new Universes which it was creating would be invisible daughters creating new life out of the death of the old.

The proposal remains a hope rather than an established theory. However if it becomes properly established, it opens the way to the concept not merely of evolution of the Universe, in the sense that its structure and contents develop in time, but in the sense that the Darwinian selection of Universe models (or rather, of expanding universe regions) could conceivably take place. The idea, proposed by Lee Smolin, is that there could be collapse to black holes followed by re-expansion, but with an alteration of the constants of physics each time, so that each time there is an expansion phase, the action of physics is a bit different. The crucial point then is that some values of the constants will lead to production of more black holes, while some will result in less. This allows for evolutionary selection favouring the expanding universe regions that produce more black holes (because of the favourable values of physical constants operative in those regions), for they will have more daughter expanding universe regions. Thus one can envisage natural selection towards those regions that produce the maximum number of black holes. The idea needs development, but is very intriguing, uniting as it does two great concepts of modern scientific understanding: the expanding universe and the evolution of populations.

These are all fascinating speculations which could conceivably be correct, but they have no experimental basis. The theory that predicts any particular one will have to be extremely convincing, if it is to be generally accepted.

5.5 Cosmic History

The overall possibilities for the Universe can be divided into four cases, as shown in Table 6.

The Universe			
Unchanging		Evolving	
Static	Steady State	Big Bang	Phoenix
Never expands or contracts	Always expanding	One expansion phase	Cycles of expansion and contraction

Table 6: Views on the evolution of the Universe

The possibility of an unchanging universe appears to be ruled out, although many people find it philosophically attractive. There may or may not be many expansion phases; this depends on the operation of physical processes that we do not understand well. However we do understand quite well much of the present expansion phase of the Universe.

> As the Universe expanded from its hot early phase, physical processes led to the production of the light elements and then of large scale structures, galaxies, stars, and the Solar system, eventually providing the habitats in which our life evolved. The Cosmic Microwave Radiation is fossil radiation left over from the hot early phase.

The physical scale of the Universe is enormous, and the images of distant objects from which we obtain our information are extremely faint. It is remarkable that we have been able to understand the Universe as well as we do.

CHAPTER SIX

EMERGING QUESTIONS
AND UNCERTAINTIES

*We have seen in the preceding chapter that as we touch the larger issues
of cosmology, we begin to reach the limits of certainty that can be achieved
by the scientific method. In this chapter we consider some of these
problems and emergent questions. This helps to put in perspective the
achievements and limits of science in understanding the material Universe.*

We consider in turn, uncertainty due to observational limits and horizons;
problems in testing the nature of fundamental forces; uncertainty about
physical origins of the universe; puzzles concerning deep connections
(Olber's paradox, Mach's principle, the arrow of time). Then we turn to the
fundamental underlying issues, problems arising from the uniqueness of
the Universe; and uncertainty at the foundations.

The issues considered here are all limits on what science can achieve
within its own domain of competence. In the next chapter we will go on
to consider some limitations on what science can achieve that result from
the limited nature of that domain.

6.1 Observational limits and horizons

Our ability directly to determine the geometry and distribution of matter
in the Universe is restricted by many observational difficulties [98], including the faintness of the images we are trying to understand. However
there are much more fundamental restrictions on what we can observe.

We can only detect distant matter by means of particles or radiation it emits
that travel to us, receiving most of our information from light[1]. There are
therefore fundamental limitations on the region of the Universe we can see,

because the radiation conveying information travels towards us at the speed of light (and any material particles travel slower than this speed[2]). As we look out to further and further distances, we are necessarily looking further and further back in time (for example the Andromeda galaxy is 1 million light years away; this means we see it as it was 1 million years ago). We are therefore seeing the sources at earlier stages in their evolution. This makes it very difficult to disentangle the effects of physical evolution of the sources observed from geometrical evolution of the Universe. This is the main reason why we are unable to tell directly from observations of the rate of change of redshift with distance if the Universe will recollapse or not.

The Particle Horizon

Furthermore, because the Universe has a finite age, light can only have travelled a finite distance since the origin of the Universe. This feature implies that we can only see out to those particles whose present-day distance corresponds to the age of the Universe; the particles beyond cannot be seen by us no matter what detectors we may use (light has not had time to travel to us from them since the creation of the Universe). The effect is the same as the horizon we see when we look at distant objects on the Earth: there are many further objects we cannot see because they lie beyond the horizon. In the case of the expanding Universe, we call the horizon separating those particles[3] that we can have seen (or indeed have had any causal contact with) from those we cannot, the *particle horizon* [31,79,98]. Actually we cannot even see as far as the particle horizon, because the Universe is opaque at early times (before decoupling), as explained in the previous chapter. In reality we can see only as far as the *visual horizon* , where the universe becomes transparent; this lies inside the particle horizon, and corresponds to looking back as far as matter that emitted the blackbody background radiation (at the time of decoupling) [31].

It is because of these limits that we are able to say very little about the Universe on scales bigger than the *Hubble* size (the distance we can have seen since the beginning of the Universe, roughly 10 thousand million light years). Thus we cannot observationally distinguish between universe models that are strictly homogeneous in the large (implying conditions are the same at a distance 1 million times the Hubble size away from us, as they are here), and those that are not. If the Universe has finite spatial sections, there are at least as many galaxies outside our view as within it; while if it has infinite spatial sections, we cannot see an infinite number of galaxies, so what we can see is an infinitely small fraction of all there is. Any statements we make about the structure of the Universe on a really large scale (that is, many times the horizon size) are strictly unverifiable.

These limitations make it very difficult to tell if an idea such as the chaotic inflationary Universe idea is a true description of the real Universe, or not. In that case, at the present time huge sections of the Universe that are nearly homogeneous (but with different expansion rates, density parameters, etc.) would be separated from each other by very inhomogeneous transition zones, but these zones would not be visible to us.

It is often stated that the inflationary Universe idea solves the horizon problem. This refers to the issue of microwave background radiation isotropy, which runs into severe causal difficulties in an ordinary (non-inflationary) universe model, for then regions that could not have been in causal contact with each other appear to be in identical physical states, because they emit radiation that we measure to be at the same temperature. These causal problems in the early Universe are solved by inflation, for there the greatly increased early expansion allows these regions causal contact [20,89-91,95]. However there are still visual horizons in these Universes[4] , so the verification problem remains.

Small Universes

There is one exception to this generally pessimistic situation. It is possible (even if the Universe is a low density Universe) that the large-scale connectivity of space could be different from what we expected, so that the Universe is in fact a *Small Universe,* spatially closed on a scale smaller than the Hubble size. Then if one could go in an arbitrary spatial direction at constant time, one would eventually end up very close to where one began (as in the case of a sphere, torus, or a Möbius strip). If this were the case we would be able to see right round the Universe several times; so we could see each galaxy (including our own) several times through images in different directions in the sky, a relatively small number of galaxies giving a very large number of images [31].

The effect is like being in a room whose walls, floor, and ceiling are all covered with mirrors: you see a huge number of images of yourself fading away into the distance in all directions. Similarly in a Small Universe, despite its small size we would see a large number of images of each galaxy fading away in an apparently infinite universe. In this case (and only in this case) there would be no visual horizon, and we could in principle determine the geometry of the whole Universe by observation, for all the matter that exists would then be accessible to our observation (in contrast to the usually considered situation, where only a small fraction of that matter can be seen). Furthermore in this case we would be able to study the history of our own Galaxy by optical observations, as we would be able to see it at different times in its history in the different images that would be visible to us.

Such Universes have all the advantages attributed by Einstein and Wheeler to a closed Universe (see above); thus it is an attractive possibility. Now it is possible we live in such a small Universe, but if this were true then proving it by observation would be difficult; and there is no solid evidence that this is indeed the case. Thus the working hypothesis is that we do not live in a small Universe, but we should keep an open mind on this matter.

Limits to verifiability

Overall, what we can say with any degree of certainty is strictly proscribed by observational limits [98]. We can in principle observationally determine (a) a great deal about the region we can observe (which lies inside the visual horizon); (b) a little about that which lies outside our visual horizon but inside the particle horizon (we might be able to tell something by use of neutrino or gravitational wave telescopes, someday when technology has developed sufficiently, but this is decades into the future); (c) nothing about that which lies beyond the particle horizon: this region is unobservable by any method. In a Small Universe there are no visual horizons, but the real Universe is probably not like that. The implication is that when our models give predictions of the nature of the Universe on a larger scale than the Hubble radius, these are strictly unverifiable, however appealing they may be.

6.2 Testing the nature of fundamental forces

In trying to understand the early Universe, we also come up against major limits in terms of our ability to test the predictions of our proposals for physical laws. Even if we could build a super-collider as large as the entire Solar System, we could not reach the kinds of energies that come into play in the very early Universe, so we cannot test the behaviour of matter under the relevant conditions [89]. This puts major limits on our ability to test whether our theories of those times are right or not. For example while it is commonly believed that inflation took place in the early Universe, we have been unable so far to detect in experiments on Earth the field responsible for inflation, and so cannot confirm that the proposal for the underlying mechanism is correct. Similarly the proposals as to how synthesis of protons from quarks took place in the early Universe cannot yet be confirmed because we have not seen the relevant particles, and measurements of the decay rate of the proton contradicts that simplest theory that could explain the proposed mechanism; we do not know which of the more complex possibilities (if any) may be correct.

Indeed the early Universe is the *only* place where some of the laws of physics come fully into play[5]; consequently the situation is the reverse of what we might hope, in that instead of being able to take known laws and use them to determine what happened in the very early Universe, we may have to proceed the other way round, regarding the early Universe as the only laboratory where those laws can be tested. This has led to an important discovery; comparison of element abundance observations with studies of nucleosynthesis in the early Universe determined that there are only three neutrino types rather than four, before this question had been tested experimentally on Earth. Results from the accelerator at CERN later confirmed this conclusion.

However this type of reasoning only works when there are a few clear-cut alternative observational predictions, and depends on the assumed cosmological conditions being correct. When we consider the really fundamental questions, whose understanding is the Holy Grail of theoretical physics, even the broad kind of approach to take is not clear. We are concerned here with the unification of our understanding of all the known forces into a single theory. In other words a ''theory of everything'', combining together the features of gravity, electromagnetism, the weak force, and the strong force in a way compatible with relativity theory and with quantum theory. Various proposals have been made [20,37], of which the most popular recent one is superstring theory [38], representing fundamental particles as string-like rather than as point particles. However this has not yet been formulated in a fully satisfactory way, and also (despite early hopes that it would turn out to be unique) turns out to be a large family of theories rather than a single theory. This kind of physics probably controls the very earliest phases of the expansion of the Universe; we can reject some of the theories on the basis of their cosmological predictions, but cannot in this way select a particular one as being correct, nor can experiments on Earth distinguish between them. We certainly cannot use this broad class of theories to determine a unique history for the very early history of the Universe. Thus the practical limits of testing of physical laws are major limitations in determining what really happened at very early times (fractions of a second after the Big Bang).

6.3 Physical origins

This is a basic problem when we consider the events which occurred at the origin of the Universe which determine the circumstances of present day existence. The Big Bang theory outlined previously makes it clear that at a very early times there must have been an epoch where the ideas of classical physics simply did not apply; Quantum Gravity (a theory unifying general relativity with quantum theory) would have been the dominant factor at these times [20,34,89]. There are a number of different theoretical approaches to this topic, none of which is wholly satisfactory, so we do not even know for sure what basic approach to use in such theories [19,20,34]; and there is no way we can test these different options by Earth-based experiments. However it is these theories that underlie what we would really like to know about the nature of the origin of the Universe.

Despite this uncertainty, we can claim that major features of quantum mechanics, such as the underlying wave-like nature of matter, must apply here also; on this basis we can make quantum cosmology models with claims correctly to represent the results of the as-yet unknown theory of quantum gravity, when applied to the very origin of the Universe.

Various such theories have been proposed to explain the origin of the

Universe in terms of quantum development from some previous state (a collapsing previous phase, a region of flat space-time, a black hole final state, some kind of 'pre-geometry') [3,6]. Such approaches can provide a whole series of alternative hypotheses for the origin of the Hot Big Bang which has led to our existence, but of course they simply postpone the ultimate issue: for we then have to ask, what was the origin of this previous phase? This remains unanswered.

The no-boundary idea

One unique and intriguing proposal side-steps this problem neatly. This is the Hartle-Hawking suggestion [3,20] that the initial state of the Universe could be a region where time did not exist: instead of three spatial dimensions and one time dimension, there were four spatial dimensions. This has a great advantage: it is then possible that there can be a Universe without a beginning, for (just as there is no boundary to the surface of the Earth at the South Pole) there is no boundary to this initial region of the Universe; it is uniform and smooth at all points. Much is made of this proposal in Hawking's book *A Brief History of Time* [92], for it does indeed describe a Universe without a beginning in the ordinary sense of the word, although time does have a beginning (where there is a transition from this strange 'Euclidean' state to a normal space-time structure)[6.] The implications of this proposal will be consideredshortly; at present the concern is three-fold, related to the testability of the underlying propositions of such a theory.

Firstly, such proposals presuppose the unravelling of some of the underlying conundrums of quantum theory that have not yet been solved in a fully satisfactory manner - (specifically, the related issues of the role of an observer in quantum theory, and what determines the collapse of the wave function, which is an essential feature of measurement in quantum theory [19,34]). These do not arise as significant problems in the context of laboratory experiments, but become substantial difficulties when quantum theory (which is usually applied to sub-microscopic systems) is applied to the Universe as a whole. Second, we certainly cannot test the Wheeler-de Witt equation which underlies quantum cosmology: we have to accept it as a huge extrapolation of existing physics, plausible because of its basis in established physical laws but untestable in its own right. Even some of the underlying concepts (such as 'the wave function of the

Universe') have a questionable status in this context for they are associated with a probabilistic interpretation which may not make sense when applied to a unique object, namely the Universe.

The issue of Initial Conditions

Thirdly, and irrespective of our resolution of the previous issues, we are tackling here the problem of *initial conditions for the Universe:* we are trying to use physical theory to describe something which happened once and only once, to for which no comparable events have ever occurred (or at least, none that are accessible to our observations). The notion of a law to describe this situation faces considerable difficulties. If a 'law' is only ever applied to one physical object, it is not clear if the usual distinction between a physical law and specific initial conditions makes sense (cf. the following section). That 'law' certainly cannot be subject to empirical test in the same way as other physical laws.

Whatever 'law' we may set up to describe this situation [20,92], we have one and only one test we can do: we can observe the existent Universe and see if it is congruent with the predictions of that 'law'. If it is, this supports that law but not decisively, for there will in general be several laws or underlying approaches that give the same result; these cannot be distinguished from each other on the basis of any experimental tests. We can obtain strong support for one particular view (such as the Hartle-Hawking 'no-boundary' proposal) only by utilising the kind of criteria for good theories that were introduced in the Chapter 2.

Whatever explanation we may give for them, unique initial conditions occurred at the origin of the Universe. They determine both the initial structure of space-time, and its matter content.

The matter we see around us today is the remnant of that initial state, after it has been processed by non-equilibrium processes in the early Universe and then in a first generation of stars, as discussed in the last chapter. Thus we understand the role of initial conditions; however this analysis does not answer the ultimate issues of origin and existence, in particular why the initial conditions had the form they did. We return to the question of ultimate causation and the issue of existence , (why does anything exist at all?) in the following chapter.

6.4 Deep connections

In developing these questions, it is important to understand the *interconnectedness* of the Universe. As well as determining the initial nature of matter and of the space-time geometry, the choice of initial conditions for the Universe profoundly affects the nature of physics in other ways. We consider here three particular examples, namely Olber's Paradox, Mach's Principal and The Arrow of Time.

Olber's Paradox

The classic illustration of this interconnectedness is known as *Olber's Paradox,* and concerns the question: why is the sky dark at night? [79,99].

The point is as follows: if we consider a simple static Universe uniformly filled with steadily shining stars, then while the radiation received per star goes down with the inverse square of the distance from the observer, the number of stars goes up with the square of the distance. When we add up the effect of all the stars, the two factors in the square of the distance cancel, and we conclude that the radiation received becomes unboundedly large as we consider the combined effects of more and more distant stars. Thus the night sky should be infinitely bright, according to this simplest model. If we allow for the fact that nearer matter interposes between the observer and more distant sources, we conclude that (because each direction eventually intersects the surface of a star, and the calculation above shows that the surface brightness of a star is independent of its distance from us) the night sky (and for that matter, the day sky) should in every direction be as bright as the surface of the Sun.

Now at first you might think the problem is simply that it would be a bit uncomfortable having a bright sky at night; we'd have to keep the curtains closed to get some sleep. Nothing could be further from the truth. If this were the case, Earth could not radiate its waste energy to the sky, which would everywhere be as hot as the surface of the Sun; consequently the Earth would heat up until it was in equilibrium with that temperature. There would be no possibility of life here (the surface of Earth would be molten rock and any organic molecules would be disintegrated by radiant energy). The dark night sky is essential to life on Earth.

Why then is the real sky dark at night? There are three factors not taken into account in this calculation. Firstly, the expansion of the Universe results in the received light from distant galaxies being redshifted; this causes a diminution in the intensity of light received (proportional to the inverse fourth power of the redshift), greatly reducing the expected radiation from distant stars. Secondly stars cannot shine for an infinite time, because they only have a finite supply of nuclear fuel; so the underlying assumption that stars can shine forever is false. The model ignored conservation of energy. Thirdly, the Universe itself has a finite age, so if we look back far enough into the past we reach an era when stars had not yet turned on; the matter at that time is dark because it has not yet formed stars, and the pre-existing background radiation is nothing other than the cosmic background radiation, which is only of sufficient density to be seen as 3 K radiation today. All three factors reflect the fact that the Universe is not in a state of equilibrium, as this simple model supposed.

Thus the simple model underlying the paradox did not take into account the real nature of the expanding Universe. This is an interesting and important result in its own right [7], but it also shows us how we cannot ignore the effect of distant matter just because it is so far from us. There is so much of it, that its effects could be very important for daily life.

Mach's Principle

Another famous example of this type concerns the origin of inertia, and is known as *Mach's Principle*. This starts with a simple fact that has puzzled physicists for 300 years: the fixed stars (in modern terminology, very distant galaxies) stay in fixed positions in the sky, when compared with a non-rotating local reference frame, defined by local dynamical experiments. Specifically, while stars appear to move across the sky relative to the (rotating) Earth, they appear fixed relative to the plane defined by a Foucault pendulum (or its modern equivalent, rapidly rotating gyroscopes, as used for the inertial guidance of submarines and aircraft). The question then is, is this rather striking fact just a coincidence, or is there some underlying causal mechanism that can explain it? [100,101].

Now *inertia* is the property whereby a freely moving body continues in a straight line relative to a non-rotating reference frame, but moves on a curved path relative to a rotating reference frame (due to 'inertial forces', such as the centrifugal force that pushes one towards the side of a car as it turns a corner). Indeed it is just the absence of 'inertial forces' in a non-rotating reference frame which defines it to be non-rotating; so a causal explanation of the puzzle just posed must relate local inertial properties to distant stars. This fits in with the ideas of General Relativity, according to which gravity (a long-range force) and inertia are closely related. Thus Mach's Principle posits that local inertial properties are determined by distant matter; just as in the case of Olber's paradox, each single star contributes very little, but there are so many of them that the total effect, taking the contribution from every star, is very large. The identity of the local inertial rest frame and the rest frame of distant stars is not a coincidence: it arises because local inertia - which underlies all local dynamics - is *caused* by distant stars.

This is a controversial proposal, and it is difficult even to phrase it in a rigorous way[8] . If it were true, we could envisage the following: suppose the entire Universe contained but one galaxy, instead of the hundred billion galaxies we can see; then the inertia of one kilogram of matter would be very much less than we now measure it to be . (So for example if a car ran into a brick wall, the damage would be much less than we presently experience on Earth). If we could slowly remove galaxies from the Universe, the inertia of matter would gradually decrease. Of course we cannot carry out such an experiment, so the issue remains unresolved: there is no way we can test to see if this is correct or not. However a related possibility is that as the Universe expands it is possible that the force of gravity gets weaker; this would result in the gravitational constant decreasing with time. Several theories have been proposed in which this is true, and the effect has been looked for experimentally. The proposal has not been confirmed; if it occurs it is below the detection threshold. However it makes a very important point: it is quite plausible that if the structure of the Universe were totally different, the locally experienced laws of physics might be quite different too.

The Arrow of Time

Perhaps the most celebrated and continuously vexing of these kinds of issues is the origin of *the arrow of time* [34,102]. It is very easy to get confused about this, for the one-directional nature of time that determines our daily lives is so deeply ingrained in our experience we find it difficult to imagine how things could be otherwise.

The problem is that the *fundamental laws of physics are time symmetric:* they run equally well forwards or backwards in time[9]. Thus the undeniable existence of an arrow of time (the one-way decay associated with entropy growth, for example) is somewhat mysterious; and the more curious feature is that while one can give arguments as to why such an arrow should exist (for example, by using statistical techniques to predict the behaviour of a gas from the forces between the individual molecules), these arguments seem to work equally well both ways: they may be taken to predict the existence of an arrow of time, but cannot tell which direction of time is the future and which is the past! Thus we know that a broken glass cannot re-assemble itself from its fragments into the whole glass, even though the fundamental laws of physics assert that this is a possibility [34].

The problem is compounded in that there are several potentially independent arrows of time (those of quantum mechanics, of thermodynamics, of electrodynamics, of evolutionary biology, for example), and one of the major questions is why they all end up consistent with each other. A vexing problem that relates to this is the question of consciousness and free-will. Assuming we really do have free will, then despite the determinism of classical laws of physics, the future is not predictable from the past[10], because human intervention can alter it in a way not predicted by the laws of physics alone. This implies an absolutely fundamental asymmetry in the workings of the biological world, which are based on the laws of physics. It may well be related to the fact that although the fundamental laws of physics are time symmetric in their classical version, quantum mechanics (in its ordinary interpretation) has a major time asymmetry in terms of collapse of the wave function [34].

There are two suggestions of an answer to this conundrum: on the one hand, *the direction of the arrow of time may be related directly to the expansion of the Universe* (which would be experienced as a contraction if time ran the other way). If so, the almost inevitable conclusion is that it would be impossible for an observer ever to see a collapsing phase of the Universe; in a Universe which according to the ordinary view, reaches a maximum size and then starts to recollapse, the direction of time would reverse then. The physical situation would actually be experienced as two expansions in the opposite directions of time, coupled by a period of indeterminacy near the maximum as the arrow of time switched direction: a conclusion so strange as to call the idea into question.

On the other hand, *the arrow of time may be determined by specific boundary conditions for local physical laws at the beginning and end of the Universe,* restricting the physically realised solutions from all possible ones to those that conform to one consistent time direction [34]. Notice here that we cannot simply say that boundary conditions at the beginning of the Universe would suffice to establish this one-way flow, for until that flow is established the beginning and end of the Universe are on an equal footing: there is no intrinsic distinction between them. Thus such conditions have to be set at both the beginning and the end of the Universe.

A key issue here is how initial conditions for some physical field are correlated with each other at the start and end of the Universe. In the past they should be uncorrelated, but in the future they should be correlated. For example [34], after a glass has fallen to the ground and broken, the pieces disperse away from where it has fallen. We cannot simply give the fragments the correct reverse velocities, so as to all come together at the right time and re-assemble the glass; the correlations required are too exact. While this time-reversed motion would certainly also be a solution to the equations, the problem is the incredible degree of coordination required to achieve this. Similar issues arise, for example, in considering why a radio transmission can only be received after it has been broadcast, and not before (this time-reversed situation being a possible solution of the time-symmetric Maxwell's equations, which determine the behaviour of the electromagnetic field). The point is that in the real world, such a solution would require exact correlations of the incoming field in the past, which are unattainable. However such correlations necessarily occur in the future all the time (the radio signal, after it has been broadcast, arrives in undistorted form in thousand of receivers; the music they all play is therefore highly correlated).

Many see this as the key feature in the arrow of time: *there are different correlations in the future and the past.* However the question arises in relation to the issue of free-will whether this is a *cause* of the arrow of time or merely a *description* of its effects. Penrose has suggested that when one takes into account the contribution of gravity to entropy growth, it is *smoothness of initial space-time structure* that is the key feature distinguishing the beginning of the Universe from irregularity and roughness that characterises its end [34]; but this view is not shared by all. An alternative view, proposed by Prigogine, is that we should aim at a reformulation of the laws of physics to incorporate the arrow of time in its very foundation, contrary to our present understanding of these laws.

Whichever kind of interpretation we may suggest, it is clear that on our present understanding of the nature of physics, the arrow of time is not embedded in the fundamental laws, but is a property of the boundary conditions for physical quantities imposed at the beginning (and probably also at the end) of time. The situation could be quite different in universes with different boundary conditions. Whatever theories we may have about this cannot be tested by any physical experiment; but the conclusion is of the utmost importance for daily life, and indeed for the very existence of life (which could not function without an arrow of time).

The unity of the Universe

Overall these examples point to deep connections and unity of the physical Universe, not merely in terms of effect of microphysical laws on macroscopic structures, as envisaged in the inflationary Universe picture, but also in terms of the very nature and functioning of those laws [101].

Indeed, the examples just given show there may be no clear-cut distinction between boundary *conditions for physical laws* at the beginning of the Universe, and *the nature of local physical laws;* for the boundary conditions for those laws at the beginning of time are given as part of the structure of the Universe, and cannot be changed; but this is the essential feature characterising the physical laws themselves. What from the viewpoint of an ensemble of Universes is just one of a whole set of possible boundary conditions, may critically affect the nature of local physics within a specific Universe in a way that is experienced as absolute and immutable, so that (in that Universe) it is indistinguishable from a immutable physical law. Thus in the cosmological context, the distinction between initial conditions and physical laws can become blurred, or at least these features may be highly interdependent. However it is these interconnections that provide the setting within which life can exist.

6.5 The Uniqueness of the Universe

What we run up against time and again is the fundamental feature of *the uniqueness of the Universe,* and the problems this gives rise to as we try to unravel its nature [103,104].

Cosmology is the ultimate historical science [11], for by definition there is only one Universe. In any other historic science there are other similar objects to compare a particular individual object with, (in geology, there are many mountains and a number of continents; in astronomy, there are numerous stars and galaxies, and many planets; in evolutionary theory, there are many different species that have related evolutionary histories). Only in the case of cosmology is there nothing whatever we can compare with the subject of study (the Universe). This is the ultimate reason why, when we penetrate to the heart of the matter - the choice of particular physical laws that govern the Universe, and of the particular initial conditions that occurred in the one unique Universe - our theories simply cannot be subject to confirmation in the normal sense.

We cannot perform the kinds of experiments that experimental sciences rely on (there is no way we can alter its initial conditions and see the resultant effects), and we cannot even do the kinds of comparisons with similar objects that underlie the other historical sciences. We can only observe what is there, and compare predictions with observations. In this way we can learn a lot about the physical nature of the Universe and the way it functions, as described in Chapter 5, as much of this is based on observation; but we run into problems when we try to answer issues of the kind considered in this chapter, particularly those related to initial conditions. In this case we include a theory in a list of possible theories if its conclusions are not in blatant contradiction to the observations. (As pointed out previously, small discrepancies can usually be explained in a myriad of ways that maintain the integrity of the main theory: the sources evolved, selection effects occurred, there was an unrecognised interfering factor, and so on.) We then choose between the theories on the basis of (non-verifiable) philosophical criteria.

The conclusion is that

we have to evaluate theories of the Universe knowing that they are testable but intrinsically unverifiable, in the sense just explained.

Because of this,

the choice of competing theories is largely dependent on the philosophical stance adopted (whether this is explicitly acknowledged or not); specifically, the crucial feature is the choice of criteria of what is a 'good' theory and what is not (as discussed in Chapter 2).

Cosmology is more dependent on such criteria than any other science precisely because of the uniqueness of its subject matter. Given a choice of such criteria, the evidence will strongly constrain what is acceptable as a theory and what is not, and may even lead almost uniquely to a specific understanding.

Are such criteria themselves subject to experimental test? To a certain extent, in that past evidence shows what has worked well in general as criteria for choosing theories in different areas of understanding; and this plays a considerable role in our choice. However cosmology is different from all other disciplines; in the end an unavoidable choice must be made that is essentially philosophical and not subject to experimental test.

It will be the contention of the rest of this book that *we should use broad criteria that take into account the whole range of human experience, and not just that part which can be scientifically described* (though that, of course , must be included as a central feature). In the remaining chapters, we first look at some of the issues that may form part of such a broad view, and then at some of the range of possible answers. Before moving on to this, a final comment on uncertainty is in order.

6.6 Uncertainty at the foundations

The reader may be beginning to be dismayed by the uncertainties that are apparent at the foundations of fundamental physics and cosmology, despite the hard-won successes of the physical sciences. To complete the picture we must note that, despite what one might think, certainty is not attainable even in the logical sciences.

At a first glance mathematics itself rests on impregnable logical foundations. However determined attempts to prove this failed, and resulted eventually in the mathematician Kurt Godel showing the impossibility of proving the consistency of mathematics [13,34]. Computer science cannot help: indeed Turing and Church have shown there is no general algorithm for deciding mathematical questions [34]. Furthermore the concepts underlying probability theory, which is required in order to test any physical theory on the basis of real (noisy) data, are also dubious, because the concept of a random number is very difficult to pin down [19].

The lack of certainty

The conclusion of the chapter is that, within its own domain, there are considerable limitations on what science can determine, in respect of verification of laws and confirmation of the nature of reality; these limits prevent us from obtaining many of the desired answers and checking the validity of our theories and models. While some are the result of practicalities and the current state of technology, ultimately some of these limits are absolute; for example (unless the Theory of Relativity is disproved some day) the speed of light is an absolute limit to communication, and consequently the limits on what we can observe in the Universe (in particular, through the existence of the particle and visual horizons) are absolute: no advances in technology will change them. Furthermore the ability of science to answer foundational questions is strictly limited.

What attitude should we take to all this? I would suggest that it confirms the profound conclusion that *certainty is unattainable at the foundations of understanding in all areas of life,* including fundamental physics and cosmology, as well as philosophy and theology; even the apparently impregnable bastion of mathematics is vulnerable to this comment. This is not the same as saying that anything goes (as some in the arts and social sciences appear to believe), but rather that what we can learn with reasonable confidence *concerning foundational issues* is strictly bounded[12].

This may seem obvious to you, and you may even find it easy to put into practice. If so you are very modern in your attitude to knowledge and are able to counter a very deep strain of thought over the past centuries. Historically, while people have from time to time shifted the focus of the hoped- for certainty (for example from theology to science), they have consistently sought for it. Many who claim to be 'rationalist' or 'free-thinking' are just as dogmatic as any fundamentalist theologian or reductionist scientist, nor are the social sciences free from dogmatic stances and closed minds.

This does not mean that we must give up the hope of attaining a good understanding of the way things work; rather it means that

> a mature attitude must take this problem of uncertainty into account, and make it a central feature of the way we approach any understanding of the Universe.

This was precisely the aim of the approach advocated in Chapter 2. Furthermore, the complexities we have run into, in terms of interdependencies and even the very notion of physical law, confirm that we need to realise explicitly that

> the models and theories on which we base our understandings are partial representations of reality, not to be confused with reality itself.

They can be very useful representations of reality in a limited domain, providing excellent understanding of that domain, but cannot rest on a foundation of absolute certainty. They cannot ever be infallible guides to reality, for they are not the same as reality.

Once we have accepted these limitations, giving up the unattainable hope of certainty, we can achieve satisfying and even profound understandings of the Universe and the way it works, provisional without doubt, but nevertheless offering a satisfactory world-view and basis for action.

PART THREE

HUMANITY AND THE UNIVERSE

CHAPTER SEVEN

THE CONGENIAL ENVIRONMENT

This chapter looks at the issue of why the Universe has the very special nature required in order that life can exist. Various basic ways to explain this fact are explored, leading to an assessment of the limits of what can be achieved by the scientific method taking into account its limited range of applicability. This completes the foundations needed in order to approach the broader domain of concern that is the subject of the next chapter.

7.1 The Anthropic question

The previous discussion has made it clear that initial conditions are such as to provide a suitable environment for the Earth, and for life on the Earth. One of the most profound, fundamental issues arising is the Anthropic question [91, 95, 105-108]:

Why have conditions in the Universe been so ordered that intelligent life can exist?

The point is that a great deal of 'fine tuning' has taken place in order that life be possible; in particular various fundamental constants need to be highly constrained in their values for life as we know it to exist. There are many relationships embedded in physical laws that are not explained by physics, but are required for life to be possible. How has it come about that the Universe permits the evolution and existence of intelligent beings at any time or place?

The issue of fine tuning.

It is easy to consider universes where life would not be possible. There could be a universe that expanded and then recollapsed with a total lifetime of only 100 thousand years; evolution could not take place on that timescale. The background radiation might never drop below 3000 K, so that matter was always ionised (electrons and nuclei always remaining separate from each other); the molecules of life could then never form. Black holes might be so common that they rapidly attracted all the matter in the universe, and there never was a stable environment in which life could develop. Cosmic rays could always be so abundant that any tentative organic structures would be destroyed before they could replicate.

There are many ways in which the boundary conditions in a universe could prevent life occurring. But additionally, we can conceive of universes where the laws of physics (and so of chemistry) were different to ours. Almost any change in these laws will prevent life as we know it from functioning. If the neutron mass were just a little less than it is, proton decay could have taken place so that no atoms were left at all. The production of carbon and oxygen in stars requires the careful setting of two different nuclear energy levels; if they were just a little different, the elements we need for life would not exist [91]. Perhaps most important of all, the chemistry on which the human body depends [24] involves intricate folding and bonding patterns that would be destroyed if the fine structure constant (which controls the nature of chemical binding) were a little bit different.

To understand the importance of this, one must appreciate the complexity of what has been achieved [49-53]. The structure and function of a single living cell is immensely complex. But a human grows to an interconnected set of 10 thousand billion cells, all working together as a single purposive and conscious organism in a hierarchically controlled way (the *organisation* issue), put together according to instructions in the DNA molecules that are read out and executed in an order that depends both on time and position (the issue of *development,*), able to function continuously all the time as the number of cells increases coherently from 1 to 10 thousand billion in a highly organised fashion, passing through different stages of maturity (the issue of *growth*), all of this happening in an interacting set of organisms of similar levels of complexity within a hospitable environment (the *ecosystem* issue), this system itself developing from a single cell to the level of complexity we see around us today (the *evolution* issue), all the while remaining functional. All of this is possible because of the nature of quantum mechanics (essentially the Schroedinger equation and the Pauli exclusion principle,) and of the forces and particles described by physics (essentially the electromagnetic force acting on the proton and the electron, together with the strong force binding the protons and neutrons in the atomic nuclei), which together control the nature of chemistry and hence of biological activity. They all fit together as required because of the precise values taken by the fundamental constants that control the strengths of physical interactions, which happen also to allow the functioning of stars as required in order to produce the needed elements, and allow development of the solar system (which is made possible through the force of gravity), with a hospitable surface for life on the Earth [91,95,107] (one of the key elements here being the remarkable properties of water [49,95], which again would be different if the fundamental constants were different).

The nature of this achievement is truly awesome. Modern moves towards determining a unified fundamental theory of all forces could make it even more amazing, because if physics ever achieved its aim of evolving a single theory with essentially no free constants [37,38], then these extraordinarily complex structures would be the result of the action of that unified theory: in effect, the nature of the unified fundamental force would be preordained to allow, or even encourage, the existence of life.

In summary, to allow life to occur, we require the existence of heavy elements; sufficient time for evolution of advanced life forms to take place; regions that are neither too hot nor too cold; restricted values of fundamental constants that control chemistry and local physics; and so on. Thus *only particular laws of physics, and particular initial conditions in the Universe, allow the existence of intelligent life.* No evolution whatever is possible if these laws and conditions do not have a restricted form. Apart from all the specific features of the laws of nature that allow complex functioning, there are four general features that are of importance. Firstly, as emphasized by P C W Davies [6], the concept of *locality* is fundamental, allowing local systems to function effectively independently of the detailed structure of the rest of the Universe. The complex of interacting systems in a human body could not possibly work if this were not so. Secondly, the *existence of an arrow of time,* and hence of laws like the second law of thermodynamics are probably necessary for evolution and for consciousness. This depends on boundary conditions at the beginning and end of the Universe (cf. the discussion in the previous chapter)[1] . Thirdly, physical conditions in our environment must be in a *quasi-equilibrium* state, or the delicate balances that permit us to exist and evolve will not hold. Finally, presumably the *emergence of a classical era* is required (the very early universe would be a domain where quantum physics would dominate, leading to complete uncertainty and an inability to predict the consequence of any initial situation; we need this to evolve to a state where classical physics leads to the properties of regularity and predictability that allow order to emerge).

All of these will occur only if the boundary conditions of the Universe are chosen in a particular way. They will not be true in a generic universe. Thus the Universe provides a hospitable environment for humanity. Why is this so? Because of the deep connections just discussed, this is not an issue related to only one aspect of the structure of the Universe. It refers to the total inter-related organisation of the laws of nature and the boundary conditions for those laws, that fashion the Universe as we know it.

The Weak Anthropic Principle

There are two purely scientific approaches to the Anthropic issue[2]. The first is the Weak Anthropic Principle (WAP), based on the comment that it is not surprising that the observed Universe admits the existence of life, for the Universe cannot be observed unless there are observers in it [95,106,107]. This seemingly empty statement gains content when we turn it round and ask, at what times and places in the Universe can life exist, and what are the inter-connections that are critical for its existence? It could not, for example, exist too early in the present expansion phase, for the night sky would then have been too hot. Indeed from this viewpoint the reason the observed night sky is dark at night is that if it were not dark, there would be no observers to see it. Furthermore one can deduce various necessary relations between fundamental quantities in order that the observers should exist (e.g. those mentioned above). If for example the fundamental constants vary with time or place in the Universe, life will only be possible in restricted regions where they take appropriate anthropic values.

Hence this view basically interprets the Anthropic Principle as a selection principle: *the necessary conditions for observers to exist restricts the times and places from which the Universe can be observed.*

This is an interesting and often illuminating viewpoint (for example, neither the Chaotic Inflationary Universe idea nor the Many-World interpretation of Quantum Cosmology [3] work unless we add such an Anthropic component into their interpretation to explain why we observe the Universe from a standpoint from which it appears homogeneous and isotropic). Its successful use may be claimed to depend on use of Bayesian statistics to underpin inductive inference [108]. However it is also a conservative approach, avoiding the main issue under discussion in this section.

The Strong Anthropic Principle

By contrast, the *Strong Anthropic Principle* (SAP) tackles the issue head on [95,106,107], claiming that *it is necessary that intelligent life exist in the Universe; the presence of life is required in order that a universe model make sense.*

Considered purely scientifically, this is clearly a very controversial claim, for it is hard to provide scientific reasons to support it. The most solid justification attempted is through the claim that existence of an observer is necessary in order that quantum theory can make sense. However this is based on one of a number of different interpretations of quantum theory; the nature of these quantum foundations is controversial [19], and probably falls within the untestable category of issues discussed above. Furthermore if we were to suppose this argument correct, then the next step is to ask, why does the Universe need quantum mechanics anyway? The argument would be complete only if we could prove that quantum mechanics was absolutely necessary for every self-consistent universe; but that line of reasoning cannot be completed at present, not least because quantum mechanics itself is not a fully self-consistent theory. (Apart from the logical issues at its foundation, it suffers from divergences that so far have proved irremediable in the sense that we can work our way round them to calculate what we need, but cannot remove them.)

Neither argument by itself gives a convincing answer to the Anthropic question. To make progress, we have seriously to consider the nature of ultimate causation.

7.2 Issues of ultimate causation

The question posed is, what is the foundational cause for the surface phenomena we see? That is, if we pursue the chain of physical cause and effect as far as we can follow it, we are still left with the question: *why did this all occur?* [1,7,8]. Whatever the reason the ultimate cause is what we are seeking when we follow the chain of causation to its conclusion.

There appear to be five main approaches to the issue [10]: Random Chance, High Probability, Necessity, Universality, and Design. We briefly consider these in turn.

1: Random chance, signifying nothing.

Conditions in the Universe just happened. Pure chance led to things being the way they are now. Probability does not apply. There is no further level of explanation that applies; searching for 'ultimate causes' has no meaning.

This is certainly logically possible, but not satisfying as an explanation except to a total reductionist (particularly because we obtain no unification of ideas or predictive power from this approach). Nevertheless some scientists implicitly or explicitly hold this view.

2: High probability.

Although the structure of the Universe appears very improbable, for various physical reasons it is in fact highly probable (the chaotic cosmology idea). These arguments are only partially successful even in their own terms, for they run into problems if we consider the full set of possibilities. (Many discussions implicitly or explicitly restrict the considered possibilities *a priori,* for otherwise it is not very likely that the Universe will be as we see it); and we do not have a proper measure to apply to the set of initial conditions, enabling us to assess these probabilities. Furthermore, application of probability arguments to the Universe itself is dubious, because the Universe is unique.

Despite these problems, this approach has considerable support in the scientific community. For example it underlies the inflationary proposal for cosmology discussed above.

3: Necessity.

Things have to be the way they are; there is no option. This can be taken in a strong form or a weak one. The strong form is the claim that the features *we see and the laws underlying them are demanded by the unity of the Universe* (coherence and consistency require that things must be the way they are; the apparent alternatives are illusory). It is really the claim that only one kind of physics is self-consistent: all logically possible universes must obey the same physics. The weak form is that only *one kind of physics is consistent with the sort of world we actually see around us.*

To prove either successfully would be a powerful argument, potentially leading to a self-consistent and complete scientific view; but we *can* imagine alternative universes! - why are they excluded? Bondi has emphasized that insofar as the view that there is a major unity underlying the Universe is valid, sufficient study of any part of Universe will reveal its whole structure, because of bonds of necessity. However a partial counter-argument is provided by considering the locality of physics, mentioned above: we are able to predict what will happen in a laboratory without knowing the total state of distant regions of the Universe. Furthermore we run here into the problem that we have not succeeded in devising a fully self-consistent view of physics: neither the foundations of quantum physics nor of mathematics are on a solid, consistent basis. Until these issues are resolved this line cannot be pursued to a successful conclusion.

In this and the last case, life exists essentially as an accidental by-product of a probable or a necessary situation; in effect, as far as the existence of life is concerned, we have a more sophisticated version of the 'chance' argument. Indeed if one could make the 'necessity' argument stick, or even determine a 'theory of everything' that explained the relations between the fundamental constants, then (as mentioned above) the issue would become even more mysterious: for why would this unique theory have precisely the qualities required to allow life? It would be a most extraordinary kind of coincidence linking the foundation of physics to emergent layers of meaning in the macroscopic world.

4: Universality.

This is the stand that 'All that is possible, happens': an ensemble of universes is realised in reality. In its full version, the Anthropic principle is realised in both its strong form (if all that is possible happens, then life must happen) and its weak form (life will only occur in some of the possibilities that are realised; these are picked out from the others by the WAP, viewed as a selection principle). There are four ways this has been pursued.

1. The view may be that this happens in *space* through random initial conditions, as in chaotic inflation. While this provides a legitimate framework for application of probability, from the viewpoint of ultimate explanation it does not really succeed, for there is still then one unique Universe whose (random) initial conditions need explanation. Initial conditions might be globally statistically homogeneous, but also there could be global gradients in some physical quantities so that the Universe is not statistically homogeneous; and these conditions might be restricted to some domain that does not allow life. It is a partial implementation of the ensemble idea; insofar as it works, it is really a variant of the 'high probability' idea mentioned above.

2. It could happen *in time* , in a universe that has many expansion phases (a Phoenix universe, cf. the previous chapter), whether this occurs globally or locally. Much the same comments apply as in the previous case.

3. Alternatively, it could occur through the existence of the Everett-Wheeler *'many worlds' of quantum cosmology,* where all possibilities occur through quantum branching[4]. This view is controversial; it is accepted by some but not all quantum theorists [19,34]. If we hold to it, we then have to explain the properties of the particular history we observe (why does our macroscopic Universe develop to have high symmetries when almost all these branchings will not?).

4. Finally they could occur as *completely disconnected universes* : there really is an ensemble of universes in which all possibilities occur without any connection with each other (cf. [109]). A problem that arises then is, what determines what is possible? (for example, what about the laws of logic themselves? Are they inviolable in considering all possibilities?) We

cannot answer, for we have no access to this multitude of postulated worlds.

Problems arise with all these approaches: on the one hand in respect of testability so that we have to query the meaningfulness of the proposals as scientific explanations; (particularly in the last case, where it is not even claimed that there is some causal connection to 'our' Universe); on the other hand, in each case in order to explain our actual observations an (Anthropic) selection element has, of necessity, to be introduced (cf. the previous section), for most of the universe(s) will not look like an isotropic model; why do we live in a region that does ? Furthermore they all contradict the Ockham's razor approach to physics: they are very uneconomical in the mode of explanation. Why does this ensemble of Universes exist? (We 'solve' one issue at the expense of envisaging an enormously more complex existential reality; ultimate explanation of this reality is even more problematic than in the case of single Universe.) Nevertheless this approach has an internal logic of its own which some find compelling.

5: Design.

The symmetries and delicate balances we observe require an extraordinary coherence of conditions and cooperation of laws and effects, suggesting that in some sense they have been purposefully designed, (i.e. they give evidence of intention, realised both in the setting of the laws of physics and in the choice of boundary conditions for the Universe[5]).

This is the basic theological view [6]. Unlike all the others, it introduces an element of meaning, of signifying something (in all the other cases, life exists by accident; as a chance by-product of processes blindly at work [65]). The prime disadvantage of this view, from the scientific viewpoint, is its lack of testable scientific consequences. (''Because God exists, I predict that the density of matter in the Universe should be x and the fine structure constant should be y'';) this is one of the reasons why scientists generally try to avoid this approach. However it is the aim of this book to look at issues from a broader stance than the strictly scientific; this will be pursued in the following chapter.

To make sense of this view, one must accept the idea of transcendence: that the Designer[6] exists in a totally different order of reality or being, not restrained within the bounds of the Universe itself. A scientific analogue of this idea is given by the now commonplace concept of the imbedding of the Universe in a higher dimensional space, where ultimate reality (the higher dimensional space) is of a different order to the reality experienced by those restricted, by their structure and sensory apparatus, to the 4-dimensional imbedded space-time. On this kind of view the Universe can be viewed from outside itself as a whole (incorporating all history: past, present and future) and creation involves the causation of this whole. The ultimate issue of existence of this higher reality (and of the Designer) remains unanswered (as it does in all approaches).

Note here that the modern version of the Design argument is different from the old one. Originally God might have been envisaged as specifically designing human beings. The more modern version, consistent with all the preceding scientific discussion, would see God designing the Laws of Nature and the Boundary conditions for the Universe, in such a way that life (and eventually humanity) would then come into existence through the operation of those laws, leading to the development of specific classes of animals through the process of evolution as evidenced in the historical record [65,67]. The issue we are concerned with, in terms of ultimate causation, is the nature of matter and fundamental forces, and why they not only admit the existence of life but even prefer it.

As discussed above, the laws of physics and chemistry are such as to allow the functioning of living cells, individuals, and ecosystems of incredible complexity and variety, and it is this that has made evolution possible. What requires explanation, is why the laws of physics are such as to allow this complex functionality to work, without which no evolution whatever would occur. We must take into account here the perilous nature of the evolutionary procedure: many false starts can occur, with entire species repeatedly being wiped out [67]; even once life has started [68], it is far from certain that it will reach the heights it has done on Earth. Nevertheless if the laws and boundary conditions are right, evolution of intelligent life is possible.

Given the acceptance of evolutionary development, it is precisely in the choice and institution of particular physical laws and initial conditions, allowing such development, that the profound creative activity takes place, and it is there that one might conceive of design taking place.

However from the standpoint of the physical sciences, there is no reason to accept this argument. Indeed from this viewpoint there is really no difference between design and chance, for they have not been shown to lead to different physical predictions.

The essential choice

For the present, the major point is that in considering the Anthropic issue, *we are faced ultimately with a choice between one of the options above as to the nature of ultimate causation of the Universe.* It must be emphasized that particular models of the nature of the physical Universe (a 'big-bang' origin, a steady state Universe without temporal origin, chaotic inflation, a Universe starting according to the Hartle-Hawking 'no-boundary' ideas) do not seriously affect the issue of fundamental causation, despite claims to the contrary; *a priori*, the existence or non-existence of a Designer could be compatible with any of these modes of realisation of a physical Universe. Indeed a Designer could choose to work through any of the other proposed 'fundamental' approaches, should he or she wish to do so.

If we look at the Anthropic Issue from a purely scientific basis, we end up without any resolution, basically because science attains reasonable certainty by limiting itself to restricted aspects of reality; even if it occasionally strays into that area, it is not designed to deal with ultimate causation. By itself, it cannot choose between these options. A broader viewpoint is required to make progress. The argument that follows in the next chapter will develop the theme, taking into account both the scientific and broader viewpoints. It will be claimed the Anthropic question can be viewed in this broader context without incompatibility, and that a far more satisfactory overall view is obtained than if we restrict our considerations to the purely scientific.

7.3 Limits of applicability of science

Before we turn to these issues, it is worth summarising the reasons why the scientific viewpoint, despite its enormous successes in terms of immediate understanding and control of the physical world (and the consequent understanding of physical reality, as outlined in the preceding chapters), cannot give all the answers we want. Within its own domain, there are all the limitations discussed in the previous chapter, but that is not the present issue; here we are concerned with the limitations of that domain.

Firstly, science is limited to its domain of application (the measurable behaviour of physical objects) and so cannot handle concepts of a quite different nature, such as the meaning of beauty, the greatness of literature, the joy of cooking, the lessons of history, or the quality of meditation. It is very powerful in its domain, but that domain is strictly limited.

The scientific understanding of complex systems governed by physical laws (as in biology) allows for emergent structure at higher levels of the hierarchical layers of organisation, incorporating new layers of meaning at the different levels of structure (cf. Chapter 4); but science does not explain such meaning in a serious sense (this theme will be discussed in the following chapter). Furthermore the limits of what will be achieved in understanding such structure at a functional level are as yet undetermined. At present science cannot cope with issues such as consciousness and freewill. It is an open question if it will ever do so; it is not clear whether they lie in its domain of explanation or not.

Secondly, underlying the physical view of the world is an layer of structure, a *Metaphysics* (lying at a deeper level than physics), that is related to the following issues:

* why does anything exist at all (why does matter exist? why does the Universe exist?)

* what underlies the nature of physical laws (why do any laws exist at all? Why do they seem to have a mathematical nature? In what way is their compulsory operation imbedded in reality?)

* what determines the nature of the specific laws that in fact govern the Universe. (Why do they have the specific form they do?)

Physics itself cannot answer any of these questions. It is easy to be misled by the descriptions of physical laws given by physics, into thinking that these explain their action in some deep sense; whereas they are in reality simply a form of naming, of labelling what happens [22]. This is helpful in understanding and modelling what occurs, but does not get at the essence of how the laws really work. (To name one interaction 'the force of gravity' helps us to predict what will happen, but does not really tell us how matter is able to exert an attractive force on other matter a long way away; relabelling it 'the effect of the gravitational field' does not alter this situation). The point is that physics can comment on physical laws actually in operation, but not on where they come from or why they exist[7]. This is obvious if you think about the scientific method and the restrictions on verifiability considered previously, (and ultimately grounded in the uniqueness of the Universe and so also of the laws of physics in the Universe). Such metaphysical issues have to be decided on the basis of criteria that lie outside the domain of physics itself.

It must be emphasized that these limitations on what science can achieve will be unaffected by scientific and technological progress.

> It is because of the very nature of science that these limits exist; they will therefore remain, irrespective of scientific advances that may be made in the future. Essentially, investigations of the foundations of science are beyond the scope of science itself.

This is why Anthropic questions cannot be decided on purely scientific grounds; they are inevitably concerned with the issues raised here, which transcend the competence of physics (or any other physical science) to answer.

> Because of these limits, the scientific method by itself can only give an inconclusive answer to the Anthropic question.

A broader approach is required to produce a satisfactory answer. We now turn to this.

CHAPTER EIGHT

THE UNIVERSE AND HUMANITY

This chapter looks at a synthesis of understanding, a Cosmology in the broad sense, comparing the two fundamental viewpoints (a materialistic view, and a broadly religious one). The case is made that when placed in a wide context which takes humanity seriously, the second viewpoint has the potential to provide the more satisfactory world view. Whether or not this is accepted, it is at least a logically coherent position.

8.1 The larger view

Our concern is to include in our deliberations the Anthropic question of the existence of life (as discussed in the previous chapter), but also the tragedy and pathos, courage and fear, hope and irony of life in its fullness [54, 110-118]. Our Cosmology must be able to take account of the magnificent gestures of humanity, and the perceived beauty of the world around. It needs a warmer understanding than is provided by physical cosmology on its own.

Does the cold logic of science rule such a broad view out? No, it can tolerate it. Is hard scientific proof available for the kind of worldview we desire? No, it is not. However the logic of those areas (psychology, the arts, sport, morality, even religion) is highly developed in its own terms. Science can attempt to say they are meaningless, or can attempt to come to terms with them in a greater synthesis, stepping out of its rigid frame of logic and proof, and the unattainable desire for a world view which is strictly falsifiable. The real issues cannot be approached using only the scientific method based on traditional scientific data; but can be by using the essence of that method in a broader context [4,5]. In doing so we go beyond the Anthropic debate [91,95,105-107] summarised in the previous Chapter: the main point being that despite the name, that argument did not really relate to humanity seriously; it is concerned with conditions for any complex structures to form, and is no more concerned specifically with mankind than with amoeba, worms or fishes. The aim here is to bring some real human concerns into the discussion.

This book so far has mainly considered the 'hard' sciences, where precise quantitative laws can often be shown to hold. Understanding is much less clear in the social sciences (anthropology, sociology, economics, political science, psychology) and practical sciences in the social area (management, psychiatry, law), where the complexity of the systems under study prevents hard predictive power. Nevertheless they can (if properly conducted) still be pursued along the lines of sceptical enquiry outlined in Chapter 2, each using the data appropriate for its own domain; and one can determine general organising concepts and broad rules of behaviour applying in each of these spheres. Together these give major paradigms of understanding, often forming an important basis for the conduct of everyday life. However one of the principal problems has been the lack of resolution of struggles between competing paradigms in most of these areas[1]. In some cases, considerations of a basically scientific nature will resolve the issue (for example, making clear the lack of validity of astrology as a method of predicting the future), but in many cases this is not so; essential issues of the humanities and social sciences are involved.

The common human condition

The viewpoint here will be that even while such disagreements remain unresolved, leading to disputes about the proper nature of analysis, there is nevertheless a layer of understanding where high levels of certainty can be reached; that indeed *there is a universal basis to much social experience, behaviour, and practice, despite the many relativities of what happens on the surface.* In order to discern this underlying universality it is necessary to *look at the situation at a broad enough level of generality.*

It is possible, for example, to say that the functioning of societies [119-121] is made possible by (formal or informal) allocation of different roles to its different members, leading to an understanding of what part each will play; that there will be some kinds of mechanisms for allocation of resources between the members of the society; and that there will be some kind of sanctions supporting compliance with specific expected standards of behaviour. The details will vary widely from society to society, but these basic principles will operate in all. Again while most but perhaps not all psychologists would agree about the existence of the unconscious and the basic mechanism of projection, it is well-established that the most important influence in the formation of personality (apart from hereditary features) is the relation of the child to its parents (or other adults who plays the role of parent), while other role models also play a major part in guiding behaviour [54,118,121]. Such insights are able to provide an understanding of much of human conduct, based on the fact that we share not only a common physical nature, but also common economic, social, and psychological needs (which may be answered in a variety of culture-dependent ways). Furthermore, it is this common humanity that speaks to us in the great literature, art, music, and theatre of all ages; it is for example the reason that the writings of Shakespeare, Dante, Tolstoy, and Dostoevsky have such a universal appeal.

Our concern here is with the broad area of purpose and meaning. For the present argument, five fundamental features are of importance in these areas.

First, there is the *drive to meaning* that is a universal feature of human life [116],manifested in rituals [111] and myths [112] that underline and reinforce the meaning seen in life by different social groups [121], leading to a universal striving for order [122]. Science comes to terms with this to some extent by noting the themes of emergent order and function that allow meaning to be incorporated in the hierarchical structures of biology [49], but this does not by itself explain the origin of meaning. Furthermore it is fundamental that this need for meaning demands *taking seriously individual lives and life events* [123], and not just a class of similar events. We can characterise the nature of births and deaths, marriages and divorces, in a general way; while this is of some significance, ultimately what concerns me is MY life, MY birth, MY death (or that of my loved ones). To some extent meaning is based in the grand themes of existence, nature, and fate, nevertheless ultimately it is based in specific events in a personal life: the rest mean nothing if it does not relate to the individual, as for example in the Blackfoot Indian affirmation, "Creation, we acknowledge your gift of this sacred being" when a newborn baby receives his or her name [111]. Any full Cosmological view must come to terms with this.

Second, and related to the first, is the necessity of a *source of values* that *is believable and so has moral power* . This is necessary not only in personal life but also in engineering, business, government: indeed all organisation involves value judgements as to what is worth doing [11]; this has major impact on our daily lives. To be effective, these moral values should be consonant with and reinforce the meaning seen in the world. Now many values are culture laden, particularly in the arts, but [124] core moral values (the central ideals that may or may not be lived up to) it can be claimed, are not relative; for example, those stating the value of individual life[2] . Perhaps the key point here is that *there is a universal capacity to recognise high moral quality,* however grudgingly, even though it may from time to time be suppressed in individuals or groups in response to their personal or group social situation, and even though people are very varied in their responses to moral quality and demands. Thus for example, the maxim ''No man has greater love than to lay down his life for a friend'' is universally understandable and demands assent by its very nature (though it is seldom acted on). Accompanying this is a basic *sense of justice* [122] which feels offended when this moral order is violated.

Third, and related to the other two, is the *drive to search for understanding and truth* which underlies our intellectual life and scientific success, and is the basis of our striving for order in our world [122]. This is needed on the one hand to validate meaning and values (for they will be worthless and discarded if they are found to conflict with perceived truth). On the other, specific values - integrity, courage, and tenacity (the determination and persistence to follow through to the end) are required to pursue the search persistently for the truth, whatever it may be. Without this drive to look for truth in all spheres of life, we will not learn or understand much, and our lives will be very limited in their scope and success.

Fourth, there is *hope* in the face of the limitations of life and the inevitability of death: ''man's being cannot be adequately understood except in connection with man's unconquerable propensity for hope in the future ... it is through hope that men overcome the difficulties of any given here and now'' (Berger, [122], p.61). This is closely linked with play and humour: ''by laughing at the imprisonment of the human spirit, humour implies that the imprisonment is not final but will be overcome'' (Berger [122], p. 70).

Finally, there is the drive for beauty and decoration that surfaces as soon as people have satisfied the basic needs of life, apparent in every culture and every age.

Cosmology: The Two Basic Viewpoints

The question that arises now is, *What is the origin of these basic features of human life?* [122,124]. A serious answer must be based on the evolutionary perspective of the development of humanity in the expanding physical Universe, taking place on the basis of fundamental physical laws as outlined in previous chapters.

There are, in the end, two great options. The first option is that there is no underlying directionality of meaning or purpose in the Universe; this is the intention of the first four approaches to fundamental causation outlined in the previous chapter. The second option is that there is an underlying structure of meaning beneath the surface appearances of reality, most easily comprehended in terms of deliberate Design. We consider these in turn.

8.2 The Materialist view

On this view (e.g. [65,125]), somehow by chance bits of matter have been fortunate in an accidental confluence of their properties (determined by the laws of physics) and a favourable environment, occurring because a great many physical factors and conditions just happened to make the situation propitious, so that they have been able by chance to accumulate layer upon layer of structure; but it all signifies nothing. Those conglomerates of dust and water that have by some accident generated a spark of consciousness by accumulating order over the millennia, now try to hide their true situation from themselves. They fill up their time by amusing one another, erecting barricades of belief and custom to create a facade of reality that protects them from their real situation [120]. They create myths of meaning and bravery to hide from the stark reality of surrounding darkness and meaninglessness; except for those few who face the reality and cry their anger to the sky in defiance of their meaningless fate, before the enveloping darkness ends this brief episode of accidental order and illusion. In whatever way it may be dressed up, this is the materialist hypothesis, implied by any view of fundamental causation that does not have a design element. It cannot explain the themes outlined in the previous section; it denies their validity.

Various proposals have been made that contradict the above implications of the materialist view; I will comment on only three (I do not believe others lead to a different outcome, see e.g. [126] for an overview.

The Marxist View

First, the *Marxist* view is on the face of it materialist, but makes major claims about right values and the good life [126]. Now as a scientific theory, Marxism is sadly lacking (indeed its historical claims have been disproved both by old evidence and recent events), but in view of its wide appeal it is worth enquiring what is the supposed ultimate source of value in this theory. As far as I can determine, there is no claim that the underlying values have developed as a result of the Darwinian evolutionary process whereby mankind has come into being [65]; rather it seems the claim is that the values underlying Marxism are intrinsically given and obviously right to those who have eyes to see, needing no further foundation or proof.

Thus from the viewpoint of this study, the Marxist source of values is the same as in the religious world-view discussed below: they are embedded in the fabric of the Universe, waiting to be apprehended by humanity (it is true their is a lively debate over the specific details of those values, but this does not query their foundational source; rather this is of the nature of a struggle to discover the true nature of reality). Thus its underlying theory of values (like that of atheistic existentialism [126], which seems to be underpinned by a similar view) appears in fact to be non-materialist.

Sociobiology

Second, the *sociobiology* view [127] suggests that there is a purely scientific basis for ethical behaviour. The aim is to explain morality in evolutionary terms [65,125] as arising in our minds from the Darwinian process of natural selection that gave rise to humanity, and to show that it is based on survival of the fittest, with no other intrinsic value. Originally this kind of view was mainly put forward to explain aggressive behaviour (cf. Lorentz's writings [126]), but it can be argued that in the same way that altruistic behaviour is sometimes seen in animals, this process could also explain our ethical understanding.

There appear to be considerable problems with the mode of explanation. Is it really plausible that natural selection acting on a population could select beings for higher morality? After all this is related to the consciousness of individuals, hardly the same as instinctive behaviour. It is fundamental to remember here that what we learn, and any consequent change in our behaviour, has no effect whatever on the genes inherited by our children. [53,65] Furthermore there is no 'moral gene'; so that even if instinctive cooperative behaviour is the result of evolution, despite selection pressures favouring those who take more than they are prepared to give [127], ethical learning of a conscious kind cannot be the subject of evolutionary development. Apart from this issue, in my view this approach by its very nature undermines the subject it is trying to explain: it cannot validate ethical behaviour.

If ethics really is a remnant of evolution, then why should it have any binding power upon us now? And if it does have a binding power, what direction does that lead us ? For if morality is really based on evolution, and has no other meaning, then surely we should abide by that logic and embark on a eugenics programme (for example, implementing euthanasia for all children of low IQ) to prevent the survival of the weakest, instead of trying to save the weak and helpless. If we insist (in the rationalist mode) that humans are simply complex machines without souls, then indeed we should just put them down without compunction once their age of usefulness is past. We do not follow this logic because we actually don't believe that this explanation provides the ultimate source of moral knowledge or ethical behaviour.

Furthermore, suppose we realise that some particular kind of altruistic behaviour is needed for the survival of the species; the next question is, why should we value a species? Why is survival valuable? The whole point is that this criterion ('valuable') is a non-scientific category[3] . The value of survival is taken as a given, its origin as a value is unexplained by the theory that is trying to explain value. Furthermore, in answer to the central question, why should we value an individual life?, a materialist view seriously based on evolution ultimately insists that an individual life is valueless in its own right, although it might have some meaning in terms of that individual's role in the survival of the species. In consequence this viewpoint certainly does not work at the highest ethical level. It cannot recognise as valuable the grand gestures, (the life of Albert Schweizer or Mother Theresa given to the poor, or the devotion of many years of effort given, as in the case of Helen Keller, to help one single deaf and dumb person to live a full life); but this is the level of moral action which by its very nature demands assent and commitment. It is self-validating across cultures.

Nor can this kind of approach, by its foundations, treat adequately concepts such as justice, the search for truth and meaning, or any of the highest soarings of the human spirit.

Relativism

Finally if we adopt a humanist position based neither on evolution nor dogma[4] the only real alternative seems to be the *relativist view* : denying any basis for morality in the fabric of the Universe and in the evolutionary process, we have to take it that *moral claims are simply conventions developed by society to regulate social behaviour, with no ultimate significance* . (For any claims I may make about it can, on this view, be claimed by you to be purely relative to my particular culture and therefore not binding on you.)

On this view, moral rules are just the rules of the game, rules of convenience; certainly such rules are not worth dying for. Now that we are fully rational, we should perhaps rationalise them by formalising them according to Game Theory, or in some other way debate a convenient set of rules for our society at the present time, without having any real binding power other than temporary convenience (there is no underlying set of values we can try to discover; and agreement across cultures that is binding on all is unattainable, for it violates the democratic right of each group to adopt whatever views they want). For example our society may value most an emotional state of "happiness", and therefore expend great resources on developing chemicals that induce a feeling of happiness; on the relativist view, this is as good an aim for society as any other. There is no valid way to criticise this aim in terms of other moral values. Similarly the implication is that there is nothing wrong with a society whose only purpose is the individual pursuit of material wealth: each person should grab what they can get, and only the minimum of social order prevents total anarchy.

The ethical options

Once again, this trivialises morality and denies its real power; in the end, it is not believable. The essential point is that we can stand back and ask,

What are the values we should live by now, and what is their ultimate base?

If we follow the chain of value to its roots, on a materialistic view either
(a) they are socially useful but essentially arbitrary (and therefore of no real power once we realise this); or
(b) they are of evolutionary origin. Again this undermines their status, for a historic base in evolution cannot be a true base for telling us what to do today (furthermore the underlying view is that the fittest should survive; if we take this seriously we should assist evolution by eliminating the weak ourselves).

Whatever we may say, we actually believe that there are intrinsic values of rightness or wrongness to which different societies aspire to greater or lesser degrees [122,124]; the details of approach may differ, but the underlying reality is there. Thus the third approach suggests that
(c) our moral and ethical strivings reflect some real structure in the nature of the Universe which is the source of intrinsic value.

This is the basic religious view, that we now consider.

8.3 The possibility of purpose

On this view, there is an underlying meaning to the Universe which we are dimly able to perceive, a higher order of structure reflected to a faint degree in what we experience as intimations of transcendence [122,128-136] and in glimmerings of rightness and wrongness [122,124]; and this is the basis of the moral order. This underlying non-physical structure is also both the source and the answer to the search for meaning and truth, without which mankind cannot live a meaningful life [116]. The basis of ethical understanding (the fundamental sense of right and wrong) is inbuilt in the structure of the Universe[5]. We experience this understanding (a concomitant of consciousness and free will) as given, self-obvious when we open ourselves to it; we do not expect it to be explicable in scientific or evolutionary terms.

This need not contradict the scientific view. There are of course long-standing tensions between science and religion. Most of them are based on misunderstandings of the nature of both disciplines, which are in fact largely compatible [4,5]; theology is like a combination of social and historical science, with data of a rather specific kind, as discussed below[6]. As has been discussed previously in this book, a purely scientific method cannot deal with the areas we are now considering; some other approach is called for, and theology has a legitimate claim to be heard as a serious basis for understanding. However from the scientific point of view there are two issues that need clearing up before we proceed further, as they account for a significant part of the antagonism towards religion felt by a considerable number of scientists.

Objections to religion

The first objection is the *arrogance* of much of organised religion, and specifically major sections of the Christian Church; particularly the claim to be sole possessors of the truth and arbiters of what is right. This contradicts everything put forward in Chapter 2 as to how the understanding of reality should be approached; it also discredits what the Church says and stands for, because in many cases what is claimed to be true or right is blatantly wrong, both historically (for example in the famous Galileo case and in the activities of the Inquisition) and currently (for example in the teachings and practice of the Catholic Church on population control and on the infallibility of the Pope, and of the Dutch Reformed Church in South Africa in its support of Apartheid and the murderous Total Onslaught ideology).

It is a fact that organised *religious sects, including the Church, are often a very poor vehicle for true religion.* Those on the scientific side have to appreciate that it is an error to throw out the baby with the bath water: just because some (or indeed many) religious organisations are patently in error does not mean that the entire religious viewpoint is necessarily empty. Those on the religious side who wish to regain credibility have to agree that many past religious claims and much religious behaviour was erroneous; the first step towards correcting this problem is to admit the error. Their need is to adopt a religious view which is compatible with the broad scientific approach (cf. Chapter 2)[7] , the first essential of which is humility (which is also one of the demands of true religion, as envisaged in this book).

The requirement is to remember (cf. Chapters 2 and 6) that reality is bigger than any image we may have of it; each model is partial and only comprehends part of the whole. This is true for the physical world around us; indeed, *different models may be contradictory and yet tell us important truths* (as in the case of the wave and particle models of light in modern physics). Given that this is true of the physical world we have available for immediate inspection, it may be expected to be much more true for any transcendent underlying reality ('God'). Important truths can come through Jewish, Moslem, Christian, Hindu, Buddhist, and indeed secular or humanistic understanding and values; depending on our background and history, one or other approach may be more illuminating for us personally.

Despite this variety of understandings, it may well be that one particular religion is capable of giving a better understanding of ultimate reality than the others, in the sense of giving an idea of a broader part of that reality than other partial views [8,] but this is clearly a matter of widely differing opinion; other views may give more accurate representations of specific aspects. Even if we believe in one particular view strongly, this needs to be seen from the perspective that any representation or human concept of the nature of God (whether embodied in some religion or not) will be partial and indicative only, and misleading in some aspects. Taking into account also the approach to knowledge advocated in this book, which implies limits on the ability of everyone to determine the nature of reality, it follows that no form of authority - any Scripture, tradition, Priesthood, creeds, or dogma - can be taken as absolute, in religion or in science; indeed in religious terms, any attempt to represent any of them as absolute or infallible is idolatry, an attempt to replace the living God by some usurper, an idol created by human minds. Equally one's own views, however deeply held, must also be regarded with caution and continually tested, in case they may be wrong. To be absolutely certain one is right is a guarantee that one is on the path to error[9], for then the protective error-correction-mechanisms cease to operate.

Proceeding in this spirit, one can maintain a religious world view compatible with the approach to knowledge outlined in Chapter 2, and free .rom the charge of arrogance. The possibility of a religious group running on these lines in the modern world is demonstrated *inter alia* by the Religious Society of Friends (the Quakers) [137,138]. This view is of course contrary to the 'creationist' view of religion, which sets up the literal words of the Bible as an ultimate arbiter even of scientific matters, rather than seeing where the evidence leads. In my view this leads to an impoverished view not only of science but also of religion, for it then cannot take seriously many of the issues addressed in this book.

The second problem is the *nature of action in the world* by *many claiming to be religious*. This raises the issue of counter-evidence: the history of organised religious bodies, and specifically the Christian Church, has very sorry aspects (consider the Crusades, the Inquisition, the persecutions of the Reformation, the Spanish in Mexico, and so on); they have often acted in direct contradiction to their proclaimed ideals, sometimes carrying out deeds of great cruelty and destructiveness (and in some cases continue to do so today)[10]. Part of the answer is to take note of what has just been said about religious institutions being poor vehicles for true religion; this counter-evidence does not necessarily disprove the viability of a religious world view. But it is important to note that while this negative evidence is certainly there, it is far from the whole story. The Christian Church and other religious organisations have their high points as well as the low, and their highest points are amongst the highest ever achieved by mankind, including both the dramatic gestures and major stands (there has been Mother Theresa and Bishop Oscar Romeiro; the central role of sections of the Church in abolishing the slave trade and fighting poverty; the fight led by Martin Luther King for freedom in the USA, and by Mahatma Gandhi in India; and so on), as well as the daily faithfulness of countless believers quietly doing the best they can to follow their God and improve the world around them.

This raises a fundamental point in looking at the logic of morality and ethics. These are areas that are not decided statistically on the basis of largest numbers. *The most important evidence may be a single event or the life of a single person,* which can convey all the information we need irrespective of the actions of thousands of others, because it provides an inspirational view of the possible and demonstrates in action values that are otherwise mere theories. That is,

as well as providing a view on ordinary every day life, ethical or religious theory has to explain the highest points of moral and spiritual life (as well as the lowest ones), because of their nature assigning a fundamental importance to these, even if there are very few of them.

We are concerned here with the key events that transform people's lives (such as St. Paul's conversion on the road to Damascus, or Viktor Frankl's experiences at Auschwitz [116]), and hence have the capability to change humanity.

The logic at work is not that of numbers but of transparent spiritual quality; and in that logic, one event (perhaps Christ's death on the Cross at Calvary) can transform the World (and count more than all the other evidence put together, in terms of changing our understanding of the Universe). In the present discussion,

> the aim is a worldview that will in some way make sense of the great gestures of humankind, the sacrifices made on behalf of others in the face of all reason.

It is precisely these events that the materialist world views discussed above cannot accommodate in a serious way.

The specific evidence

With the preliminaries out of the way, we can turn to considering how to proceed. What do we need to explain? The aim is to give some account of the universal features mentioned in the previous section (in particular, the basis of moral behaviour and meaning, the search for truth) together with what has been discussed above (a view on the highest levels of behaviour we have seen, as well as some account of the worst). The broad approach should be as in Chapter 2. What will be taken as tests of the truth proposed? The primary test is the religious maxim, ''By their fruits ye shall know them''; that is, although overall satisfactoriness as an explanatory world view is of importance, we ultimately test a religious world view by its outcome in the world. While understanding the fallibility of humanity and the difficulties in following a transcendent vision, we nevertheless demand of a religion that at least some of its followers demonstrate in their lives the kind of qualities they claim to stand for, acting as a beacon and a light that can be followed. As we have seen, even one person demonstrating the reality of that vision in their lives can be completely convincing.

Within the broad range of evidence adduced, there is a specific strand that needs particular consideration. This is the *mystic consciousness* and *mystical experiences* [128-130], that awareness of transcendence and ultimate reality, that may be experienced as a still small voice [137,138] or an urgent overwhelming presence [128,134-136]. I am not here referring to what have been termed paranormal or psychic phenomena. It is just conceivable that some of these have a basis in reality, despite the charlatanism prevalent in this area; however the point is that these do not usually relate to morality and rightness of behaviour. Rather I am referring to a considerable body of experience relating to *conviction of an ultimate reality that makes serious moral demands on behaviour, and indeed on the total conduct of a person's life*.

In evaluating this evidence[11], I am acutely aware of the problem of hallucination or self deception, which might explain a considerable part of these experiences. However in my view it would be a mistake to dismiss them entirely because of this possibility; we have to take seriously claims of this kind not merely because they are made persistently in a great variety of circumstances and cultures, but because (whatever their cause) they demonstrably have the ability to alter people's behaviour decisively and permanently.

As we are concerned with moral behaviour, the effect of mystic vision or religious conversion cannot be ignored. These experiences may be illusory but if so, how to explain events that deserve a serious explanation? For example (in the case of Christianity), the historical fact that a carpenter and a group of fishermen and friends were transformed by a vision to the point of sacrificing their lives for it. It is at least worth exploring the possibility that these transforming visions are precisely what they claim to be: indications of a transcendent reality underlying the surface appearances of the Universe (these surface appearances being highly misleading even as regards physical reality, as emphasized by Eddington [29]).

Fundamental Themes

Suppose, then, we take seriously the writings of the mystics and those who are commonly agreed to be spiritual leaders, in old-fashioned parlance the Saints through the ages[12] . Can we extract from their experiences and teachings some conclusions that are universal in nature but also can give us helpful guidance? I suggest there are two we can add to our list of evidence to be accounted for.

The first is the major theme of a sense of *transcendent reality underlying the Universe,* already touched on. Of course modern science agrees in the sense that it supposes transcendent laws of physics underlying the Universe, but here we mean in addition a reality that pertains to meaning and morality, with a quality that inspires awe and even reverence when it is truly comprehended.

The second is a theme that transcends time and place and culture, namely *giving up what we value, letting go of what we want,* as summarised in the paradoxical statement ''He who would save his life will lose it, he who gives up his life will save it''. It entails learning to give up that which we desire to cling to, accepting the implied loss as the basis of greater good, as summarised by Robert Bellah [133]:

> 'For the deepest truth I have discovered, if one accepts the loss, if one gives up clinging to what is irretrievably gone, then the nothing that is left is not barren but is enormously fruitful. Everything that one has lost comes flooding back again out of the darkness, and ones relation to it is new: free and unclinging, but the richness of the nothing contains more, it is the all possible which is the spring of freedom.'

This leads to a profound view of how to live at all levels of life personally and in community [54,116-118]. It is also the basis of the openness and humility that underlies the ability to learn, for that requires the giving up of preconceptions, and the refusal to claim that one is necessarily right

[11]; indeed it is the basis of the openness needed for the approach advocated in this book (based on Chapter 2 and Section 6.6).

This principle applies particularly in our relations to others, where such giving up of our own desires enables a loving response to their true needs (rather than our own imposed perceptions of their needs), and is indeed required for such a truly loving response [54]. It is central to Buddhism as practised by Ghandi, is found in Judaism, Sufism, and the Bahai faith, and is central to Christianity. It leads on to the profound concept of *voluntary sacrifice on behalf of others,* a central motif not only in spiritual writings but also in morality through the centuries: for *it is the only workable mode of eliciting a free response of love from individuals with free will* [139]. On a simple evolutionary view this is a futile and meaningless gesture, running against the millennia of struggle for existence that have brought us to where we are. From a religious perspective, this is the highest form of action humanity can undertake [140,141], and indeed on this view this kind of loving action is the ultimate purpose of our lives. This is obviously and overtly worthwhile in its own right; no further evidence of its value is needed.

If we take this seriously as something real, and not just a delusion of despairing minds, we must view it as central feature of the cosmos, rather than an accidental by-product of all the forces and confluences at work. It must in some sense represent the *nature of the ultimate reality that underlies the physical Universe[13]*; in theological terms, *the purpose of the Creator.* The problem is then to incorporate this understanding in our Cosmology.

Before doing so, I must briefly consider the charge that this view is invalid because it commits the 'naturalistic fallacy' of philosophy [8]: that is, it cannot be true because 'Indicatives cannot lead to Imperatives, and vice versa', where *Indicatives* are factual statements about how things are, and *Imperatives* are statements about how things should be. Two points are in order. Firstly, this 'fallacy' is the subject of considerable dispute; it can be shown to hold if morality is simply a convention, but not otherwise; and in this book I take the latter viewpoint, thus denying the basis of the 'fallacy'. Secondly, irrespective of the previous argument, it is clear that I can indeed, on the basis of moral leanings, set up institutions (such as the Red Cross) to carry out good works, with the aim of actually influencing the course of affairs in the world. Then an Imperative (my intentions) lead to an Indicative (the institution) which in turn enables implementation of my intentions (Imperatives). This is a close analogy to the situation I have in mind for Cosmology, and shows that whether or not the idea proposed is a true description of the real universe, at least it makes sense as a concept; no logical fallacy is being committed.

8.4 A combined view

The aim now is an analysis of the Anthropic principle, in relation to the organising principles just discussed. Thus the key idea is

> we suppose that the fundamental structure of the Universe is constructed specifically to support and make possible moral behaviour, and in particular sacrificial response, by sentient beings with free will; indeed this may be said to be the purpose of the Universe.

The task is to pursue the implications that then follow[14].

From a materialistic viewpoint, the proposal is a grossly anthropocentric view of the nature of the Universe, for - materialistically considered - intelligent beings are an insignificant feature in the vast realms of space filled with vast numbers of galaxies, each made of vast numbers of stars and clouds of dust and gas. However the religious view will be that it is precisely the highest levels of order in the known Universe (the extraordinary structure and function of the human mind, and of any other moral beings that may exist in the Universe), making possible moral and religious understanding and response, that gives the whole its ultimate meaning and indeed its rationale for existence.

This does not necessarily mean that the whole enterprise is concentrated on humanity specifically (that is, on the species that has evolved on the particular planet Earth in our particular Galaxy); but rather that it has its meaning through laws of physics and chemistry that will allow evolution of intelligent and responsive life in many places in the Universe. If we accept the pessimistic assumptions of Barrow and Tipler [95], there is one advanced civilisation in our galaxy, and so about 100 billion in the observable region of the Universe; it is possible there are many more. It is the total moral, ethical and religious response of all these beings that is seen as the underlying purpose of the whole. Thus the viewpoint is not anthropocentric in a Copernican sense of being limited to our particular existence alone, on the planet Earth; but incorporates that existence - certainly taken as being valuable and of significance - into a broader and more democratic view of the value of all intelligent life in the Universe (which almost certainly exists in many places as well as the Earth).

This view gives a fundamental role to *human creativity,* for creative response in the face of the situations challenging us is a central feature of true morality (which rejects any formula solutions to ethical problems, but rather demands personal responsible action). The challenge is to transform each situation to one where the right kind of spirit comes to dominate; to do so, each situation requires an individual response. It is probable that the creation of a mental environment with the possibility of creative response to moral challenges, necessarily leads to the universal features outlined at the beginning of this chapter.

Given the hypothesis of this fundamental aim underlying the Universe (most simply attributed to the basic nature of a Creator, otherwise supposed somehow to be embodied in the structure of reality through some kind of 'Life-Force'), we aim to examine its implications for the creation process. There are five major points [144].

1: The ordered Universe

First, there is a need for the creation of a Universe where ordered patterns of behaviour exist, for without this, free will (if it can be attained) cannot function sensibly. If there were no rules or reliable patterns of behaviour governing the activity of natural phenomena, it would not be possible to have a meaningful moral response to the happenings around one. Thus the material world, through which sentient beings are to be realised, needs to be governed by repeatable and understandable patterns of structure.

One way of attaining this is through physical laws as we know them. A very difficult question is whether this is the only way to attain such repeatable patterns of behaviour; the answer is not clear. This is partly because we do not understand the underlying basis of physical laws, in the sense of knowing how the behaviour they characterise is imbedded in matter (is there in some sense a mathematical formulation of these laws embodied in reality? Is there some kind of template for each kind of particle, embodying its physical behaviour but not described in a mathematical sense? Is the behaviour the result of the Creator simply imaging the desired results, and requiring the realised structure to conform?) However it seems likely that in whatever way they are realised in practice, they will indeed be experienced as 'laws' providing the regularity required.

For want of a better way of understanding them, I will assume that the desired regularities are indeed attained by setting particular laws of physics in action and then letting them run their course; more accurately, sustaining them so that they remain in action for the life of the Universe. Without this, everything would be chaotic and formless. Thus we may envisage the transcendent Creator at all times maintaining the nature of the physical world so that a chosen set of laws of physics govern its evolution. It must be emphasized that once this choice has been made, providing it is adhered to (as will be assumed), then the manner of action of laws will be seen by us as absolute, rigorously determining the behaviour of matter. We can then act freely within the confines of the laws, but the laws themselves cannot be altered by any action of humankind.

2: The Anthropic Universe: free will

However, we require much more than this: we require that these laws and regularities allow the existence of intelligent human beings, who can sense and react in a conscious way, and who furthermore have effective free will. The word 'effective' here means that whatever the underlying mechanisms governing human life, there must be a meaningful freedom of choice which can be exercised in a responsible way (for without this, the concept of ethics is meaningless).

We touch here on issues that science has not seriously begun to address. We do not properly understand the nature of consciousness [146-149], nor whether the 'free will' we experience is real (albeit constrained by many psychological and behavioural factors), or illusory. Some scientists maintain that it is indeed illusory, but in practice do not behave as if this were true (they are liable, for example, to talk about the difference between responsible and irresponsible behaviour - a distinction which is meaningless if free will is not real). Truly to believe that free will is not real would undermine the basis of moral behaviour, and indeed of any real meaning to our existence; a standpoint not taken seriously in practice, irrespective of the theoretical views proposed. It can be maintained that evolution can lead to consciousness in a straightforward way [149]; this still leaves open the questions raised previously in this Chapter regarding the nature of that consciousness (the basis of universal features of human concern).

If we consider, then, the implementation of laws leading to beings with free will, we are implying acceptance of the conclusions resulting from the Anthropic discussion of the previous chapters. The Strong Anthropic Principle would in effect be realised, but without the need for a basis in quantum theory; and necessarily all the restrictions implied by the Weak Anthropic Principle as conditions for the existence of life (for example, restrictions on the nature of physical laws and limits on the value of the fine structure constant) must be fulfilled. One solution to the problem of creation of living beings on the basis of physical laws is the expanding universe, with creation of structures through non-equilibrium phases, as outlined in previous chapters; an intriguing question is what other solutions might be possible. The strictly scientific criteria of the W.A.P. come into play, and give observable statements on the nature of the physical Universe; (for example, that there must be times and places where the background temperature is lower than 3000K, above which temperature a living being's body could not function, for all atoms would be ionised).

But more is implied, for we need to ensure *the conditions required to attain free will* (not normally emphasized in discussions of the Anthropic principle). As just stated, we do not know what these conditions are, except of course that they are compatible with the laws of physics and chemistry as we experience them [146]. Nevertheless it seems probable that fixed laws of behaviour of matter, independent of interference by a Creator or any other agency, is a requisite basis for existence of independent beings able to exercise free will; for they make possible meaningful, complex organised, activity without outside interference (providing a determinate frame within which definite local causal relations are possible). Thus we might envisage the Creator choosing such a framework for the Universe (thus giving up all the other possibilities allowed by the power available to Him, for example the power directly to intervene in events in a forceful way from time to time). This voluntary restriction on the nature of His creation makes possible the other major desired features, as we shall shortly see. From this viewpoint, fine-tuning is no longer regarded as evidence for a Designer, but rather is seen as a consequence of the complexity of aim of a Designer whose existence we are assuming; it is not plausible that this complexity of function can be achieved (within the context of established physical laws of behaviour) without fine-tuning.

To succeed in this all is a considerable achievement. Overall, we end up agreeing with Leibniz: the Universe is the "Best of all possible worlds" in that it is "simplest in hypothesis and richest in phenomena" (modern physics agrees with this view, in that a few fundamental laws lead to the awesome complexity of the biological world). As summarised by George Gale, "What God chose was a definition of perfection, which then functioned as a criterion of world selection".

A fundamental question following from this is, *are the features of pain and evil implied in every Universe that allows free will, as outlined here?* Almost certainly, the answer is Yes - because of the very nature of free will; for any restrictions on the natural order that prevented that self-centred and selfish use of will which is the foundation of evil action, would simultaneously destroy the possibility of free response and loving action [139].

3: The provident Universe

Given the existence of creatures with free will, one can still imagine universes arranged so that this will is constrained in an essentially unfree way, contrary to the spirit set forth in the last section. The same essential nature needs to be built into the creation of the Universe, for otherwise a free response would not be possible.

This is achieved by the impartial operation of the laws of physics, chemistry, and biology, offering to each person alike the bounty of nature irrespective of their beliefs or moral condition. The major requirement in order that this succeed is that the laws of physics do indeed allow the growth of food and provision of wealth for all humankind, which is part of the basic anthropic presupposition; (actually humankind could not evolve otherwise, so the very existence of humanity, in a world governed by physical laws, guarantees this requirement). Then rain falls alike on believer and unbeliever, and makes their existence equally possible (as opposed to a Universe where, for example, rain only falls if you praise God - presumably a perfectly possible arrangement). This mode of operation of the physical world thus fulfils the condition of freeing people from a need for obedience to the Creator in order to survive, and so makes a free and unconstrained response possible.

4: The hidden nature

A further requirement that must be satisfied to enable a truly free response, is that the created world be not dominated by the demands of the underlying morality in an unavoidable way. This excludes a creator God who strides the world, demanding obedience on pain of punishment (as in the myths of some religions). Similarly it excludes a world so dominated by explicit marks of His activity that belief in His existence and nature would be forced on everyone - with inevitable behavioural consequences [139]. This would be true, for example, if there was some mechanism that immediately punished malefactors with dire physical retribution the moment they did something wrong (your hand falls off if you steal, for example).

It is true that direct (undeniable and obvious) evidence of the existence and nature of the moral order would not by itself remove free will; but it would make demands on people's behaviour in such a way that the only options left would be either forced acceptance, or rebellion, as opposed to discernment of true choice and consequent free response. Thus the further requirement, if the Creator is neither to force humankind into direct allegiance (as a tyrant lord), nor to remove all leeway in terms of choice through the making of major behavioural demands that leave no room for doubt (and so inviting a grudging allegiance), is that the nature of underlying purpose and creative activity be largely hidden, so that doubt is possible.

This requirement is satisfied through the nature of creation as we see it, governed by impartial physical laws, which nevertheless allow a free and open response to those indicators as to creation's true nature that are given us. The ability to see the truth is dependent on readiness to listen and openness to the message.

5: The possibility of revelation

This leads to the final requirement: that despite the hidden nature of the underlying reality, it should still be open to those who wish to do so, to discern this nature, and to receive encouragement to follow the desired way. There is a two-fold requirement: firstly, that *it should be possible to make specific intimations of this reality available to those who are ready to receive them; and secondly, that there should be available to all, as a basis for ethics, a mode of revelation of a broadly based appreciation of what is right and wrong, of what is good and bad.*

In whatever way it is done, the feature I will assume is that there is indeed a channel for visions of ultimate reality[15] available to those open to them: allowing the nature of that transcendent reality partially to shine through into the reality of the world, making available to us patterns of understanding, and providing encouragement and strength to follow these visions. This feature corresponds both to the broad range of mystical experiences discussed previously [128-130], and to the specific intimations of morality and meaning that are available to us [122,124] 16 . This is similar to some degree to the way that logical and physical laws underlie the nature of Creation: they are there, intrinsic in the structure around us and underpinning physical reality, but they are not immediately obvious: they have to be discovered by those with the ability to understand them.

The physical Universe

It should be noted that none of this is in contradiction to standard physical understanding of the Universe. Rather what we have is an extra layer of explanation offered for the physical world we see around us, not in any way contradictory to physical understanding, but rather providing a kind of rationale for the need for such a reality, that is a *meta-physical explanation for the existence and nature of physical laws.*

The straightforward physicist can simply claim that there is no need for this extra layer of explanation in order to understand the physical world, and he will be right - provided we attempt to explain only physical reality, accepting without question the given nature of physical laws, and ignoring all the issues raised by the existence of a moral and ethical order (and indeed also the aesthetic dimension to life encompassed in great art and music). When we try to make sense of these extra dimensions of existence, the simple physical explanation is woefully lacking; something like that offered here is much more profound and satisfying. This view is not falsifiable in a scientific sense, but is supported in terms of patterns of evidence discussed above, and the understanding supported by that evidence (cf. Chapter 2).

We end up with a very Platonic view: there are eternal features ('forms') underlying reality. Some of them are experienced as laws of physics, somehow imbedded in the nature of matter (we need to remember the mysterious nature of how this is done, and that naming things, making models of them, and understanding the formulae to the extent of making reliable predictions, does not explain the reality itself [22]). The issue then is, what is the total extent of such underlying reality? The suggestion offered is that:

> The underlying order of the Universe is broader then described by the understandings of physics alone, and relates to the full depth of human experience, in particular providing a foundation for morality and meaning.

Within this broad framework, many detailed views of the nature of this underlying reality can be suggested; broadly they have all a religious nature. This purpose at the foundation of reality would usually be seen as in some sense a realisation of Design.

CHAPTER NINE

CONCLUSION

This chapter summarises the various alternative fundamental explanations, and assesses the proposal for a unified view stated in the last chapter.

Those who wish to, can reject the whole of the last chapter as valueless. However in doing so they will also be rejecting the possibility of relating Cosmology to some of humanity's most profound experiences. Just as Newton's laws are perfectly correct up to a boundary, but not beyond, (because their range of application is strictly limited), similarly it may be that the secular approach is fine up to a boundary; but in exploring its limitations, it suddenly becomes clear that there is more to creation than spelt out by science itself. An open-minded approach will at least consider if the broader picture might not be correct. We have now assembled all the pieces necessary to attempt some kind of assessment of the various possible ways of understanding Cosmology in the broad sense.

9.1 A deeper synthesis

Because life is the ground of consciousness, a truly remarkable feature of the Universe [82,146], it seems to require special explanation. In looking at this, it is important to follow carefully the intricate chain of cause and effect that allows us to exist and to function, reflecting a major amount of fine-tuning of physical laws [91,95]; and take into account the fact that no approach satisfactorily solves the ultimate mystery of existence [11]. Given all this, what ultimate cause can lie behind the existence of life and consciousness (the anthropic issues, cf. Chapter 7) ?

We cannot at present satisfactorily complete the argument of *Necessity*, or even of *High Probability*. If we could, the issues raised in the last chapter would still remain unanswered.

Pure chance is very difficult to sustain, despite its logical unassailability, for it seems a totally inadequate explanation of the complexity we see (it must be emphasized that the order of complexity in life is totally different from that in, say, a crystal structure, a rock, a mountain range, or a fire); nor can it satisfactorily explain morality and the desire for meaning, or the other the significant features discussed in the last chapter.

Some form of *Universality* is a possible 'ultimate' view, although it could be argued that this is really just a more sophisticated version of pure chance and suffers from severe problems of confirmation. It also provides no real guidance on moral issues. Logically it appears to include within its compass many universes with creator Gods.

Comparing the different possibilities, it is difficult to avoid the conclusion that the *Design* concept is one of the most satisfactory overall approaches when broadly conceived, necessarily taking us outside the strictly scientific arena. (This is not surprising: the scientific approach is strictly limited both as regards verifiability and its domain of application.)

Indeed the Anthropic question is where the design concept works most strongly provided we consider the context as not merely related to physics and chemistry, but to the full nature of our existence. This must include our fears and hopes, love and caring, value judgements, ethical choices and moral responsibility, as well as pain and suffering, whose reality I take to be at least as indisputable as any other area of experience. From the religious viewpoint, this explanation offers an answer to the issue of the existence of the Universe and the nature of physical laws by the highly controversial but clear statement that,

> the Universe exists in order that humankind (or at least ethically aware self-conscious beings) can exist.

The further question of 'why'? is answered by the radical claim that

> this is done so that unselfishness and love may make itself manifest, an obviously and patently worthwhile purpose.

It may or may not be persuasive, but at least it is a rational alternative that does not violate our scientific or other understanding, and indeed may be said to provide a deeper and more encompassing view of the whole range of our experience than any of the 'rationalist' alternatives. Whether one accepts this view or not, to take it seriously certainly does not involve intellectual suicide; only a closed mind, hemmed in by preconceptions and prejudice, would refuse point blank even to consider such a possibility.

Those who take the struggles and strivings of humanity to heart must take this kind of possibility very seriously, both in terms of making sense of their own situation, and in view of the many people who have espoused such views at great cost, even of their lives. A purely physical cosmology can make no sense of this; it can only account for a very narrow view of the totality of reality, in contrast to the claims of a broad-based Cosmology as sketched out above.

Those who reject this kind of view but claim an all-embracing account of the Universe (a 'theory of everything' or an interpretation of 'The Mind of God'), have before them the task of providing some other Cosmology that gives a satisfactory view of the whole of the experience of humanity, and not just the abstractions of science that 'explain' (in an illuminating and fascinating way) the functioning of the physical world. No such alternative is at present available.

9.2 Practical Implications

If one accepts a view such as that proposed above, it has practical implications as to how one should behave, for it includes a theory of ethical behaviour (albeit only sketched out here in the broadest brushstrokes). That view is compatible with the deep-rooted proposals for individual and community ethics based on the idea of 'letting go' of personal needs, and sacrifice on behalf of others [54,117]; also with ecological caring, based on a long-term view of our origin and the limits that must inform sustainable environmental and resource policy [62,63].

Its ethical and religious dimensions involve taking the whole seriously and acting accordingly [132,138]. Ultimately they involve a willingness to make sacrifices on behalf of others, even if they are not relatives or friends; indeed using this approach particularly if they are enemies, thus totally transforming the situation [140,141,150].

There is no space to pursue these issues further here, other than to remark that:

> the true ground for action of a high level is belief in a meaningful Universe with an underlying purpose and moral order. Without this, the highest aims have no basis and will ultimately falter.

It is possible that these concepts are pure illusion; that for example morality is simply a matter of convenience (and therefore has no real binding power). The meaning that people search for may be mirage. However it is also possible that the sought for meaning and morality may be real, reflecting the true nature of the foundations of the Universe.

This book has attempted to outline a coherent view of this kind which, while going considerably beyond a purely scientific framework, is fully in accord with modern scientific understanding. It may or may not be correct; I have argued that there is considerable evidence in its favour. In any case, it is at least worth consideration.

REFERENCES

Excellent references or various of the themes developed in each chapter of this book are listed below . More technical (but still accessible) works are marked with an asterisk (*).

Chapter 1: The Big Questions

Major issues in Philosophy are discussed in
[1] R C Solomon: *The Big Questions. A Short introduction to philosophy.* (Harcourt Brace Jovanovich, 1982).
The humanist reaction against the pure scientific approach is presented in
[2] B Appleyard: *Understanding the Present: Science and the Soul of Modern Man.* (Picador, 1992).
The theme of Science and Religion is presented in a modern setting in
[3] R J Russell W J Stoeger and G V Coyne: *Physics Philosophy and Theology. A Common Quest for Understanding.* (Rome: Vatican Observatory, 1989) *.
[4] J. Polkinghorne: *Science and Creation.* (Boston: New Science Library, 1988);
I G Barbour: *Religion in the Age of Science.* (Harper and Row, 1990)
[5] N Murphy: *Theology in the Age of Scientific Reasoning. Cornell Studies in Philosophy and Religion* (Ithaca: Cornell University Press, 1990) * .
[6] R J Russell N Murphy and C J Isham (Ed.): *Quantum Cosmology and the Laws of Nature: Scientific Perspectives on Divine Action.* Vatican Observatory/CTNS Conference proceedings (to be published, 1993)*

Chapter 2: Understanding the Universe

Philosophical analysis is discussed in
[7] T Nagel: *What Does it All Mean? A very short introduction to philosophy.* (Oxford University Press, 1987)
[8] J Hospers: *An Introduction to Philosophical Analysis.* (Routledge and Keegan Paul, 1959) *
(and see also [1]).

Philosophy and science is discussed in

[9] W H Newton-Smith, *The Rationality of Science* (Routledge, 1991)*
with the special issue of philosophy and cosmology presented in
[10] G F R Ellis: *Major Themes in the Relation of Philosophy and Cosmology* (Memoirs Italian Astronomical Society, 62 553-605 (1991).
The learning cycle, and organisational structure conducive to learning, is discussed in
[11] G F R Ellis: *Organisation and Administration in a Democratic Era* (Book draft, University of Cape Town, 1991).
The nature of mathematics and mathematical descriptions of nature is discussed in
[12] L A Steen (Ed.): *Mathematics Today* (Vintage Books, 1980),
[13] M Kline *Mathematics: The Loss of Certainty* (Oxford University Press, 1980)*.
Use of Bayesian statistics in analysing theories is presented in
[14] H Jeffreys: *The Theory of Probability*. (Oxford University Press, 1939)*
[15] R T Cox: *American Journal of Physics* 14 1 (1946)*
[16] A J M Garrett: Ockham's Razor. *Physics World* (May 1991), 39-42.

Chapter 3: The Physical Fundamentals

The relative scales (and so layers of order) of physical objects in the universe is illustrated in
[17] P Morrison and P Morrison: *Powers of Ten*. (Scientific American, 1982).
The basic forms of matter (solid, liquid, gas) and the nature of the elements is presented in
[18] R E Lapp: *Matter*. (Time-Life Pocket Books, 1969).
The fundamental constituents of matter are discussed in
[19] H R Pagels: *The Cosmic Code* (Penguin, 1982)
[20] P C W Davies (Ed.): *The New Physics* (Cambridge: Cambridge University Press, 1989)*.
The nature of Chemistry is surveyed in
[21] P Atkins: *General Chemistry* (Scientific American, 1989)*
[22] A Scott: *Molecular Machinery*. (Basil Blackwell, 1989),
and the formidable chemistry of life in [46] and
[23] Hanawalt and Haynes: *The Chemical Basis of Life* (Scientific American readings, 1973),

[24] A Scott: Vital Principles: *The Molecular Mechanisms of life*. (Basil Blackwell, Oxford, 1988).

The nature of Physics is presented in

[25] A S Eddington: *The Nature of the Physical World*. (Cambridge University Press, 1928)

[26] R Feynman: *The Character of Physical Law*. (M.I.T. Press, 1990),

[27] J J Wellington: *Physics for All* (Stanley Thomes, 1988),

[28] L N Cooper: *An Introduction to the Meaning and Structure of Physics* (Harper and Row, 1968)*.

The nature of relativity theory is presented in

[29] A S Eddington: *Space, Time, Gravitation: An Outline of General Relativity Theory*. (Cambridge University Press, 1920)

[30] B K Ridley: *Time Space and Things* (Cambridge University Press 1986).

[31] G F R Ellis and R M Williams: *Flat and Curved Space-Times*. (Oxford: Oxford University Press, 1989)*,

while quantum theory is surveyed in [19] and

[32] T Hey and P Walters: *The Quantum Universe* (Cambridge University Press, 1987),

[33] R Feynman: QED: *The strange Theory of Light and Matter*. (Penguin Books, 1990),

[34] R Penrose: *The Emperor's New Mind*. (Oxford: Oxford University Press, 1989).

Energy and Entropy are surveyed in [25-28,34] and

[35] *Energy and Power*. (Scientific American, 1971)

[36] P W Atkins: *The Second Law*. (Scientific American Book, 1984).

Modern developments in fundamental physics are surveyed in [19,20] and

[37] J Barrow: *Theories of Everything: The Search for Ultimate Explanation*. (Oxford Unuiversity Press, 1991),

[38] P C W Davies and J Brown: *Superstrings: A Theory of Everything?* (Cambridge, 1989).

Chapter 4: The Structures Around Us

The Applications of physics to the daily world is discussed in

[39] David MacCauley: *The Way Things Work*. (Dorling Kindersley, 1990)

[40] K Schmidt-Nielsen: *How Animals Work*. (Cambridge University Press, 1972)

and the way it underlies the development of modern society in

[41] J Bronowski *The Ascent of Man*. (BBC, London, 1976).

The implications of energy and entropy limits are discussed in [35,36] and

[42] P Colinvaux: *Why Big Fierce Animals are Rare: An ecologist's perspective*. (Princeton University Press, 1979)

[43] W F Baxter: *People or Penguins: the Case for Optimal Pollution*. (Columbia University Press, 1974)

[44] Georgescu Roegen: *The Entropy Law and the Economic Process*. (Harvard University Press, 1976)

[45] H E Daly: The economic growth debate: what some economists have learned but many have not, *Journal of Environmental Economics and Management* 14 323-336 (1987).

The issue of Hierarchical organisation is presented in

[46] Stafford Beer: *Platform for Change*. (Wiley, 1978)

[47] Stafford Beer: *Brain of the Firm*. (Wiley, 1981)

[48] R L Flood and E R Carson: *Dealing with Complexity*. (Plenum Press, 1990).

Biology is discussed in

[49] N. A. Campbell: *Biology*. (Benjamin Cummings, 1990),

(a magnificent general text), and specific aspects in

[50] C Grobstein: *The Strategy of Life*. (W H Freeman, 1974)

[51] F C Steward: *Plants at Work*. (Addison Wesley, 1965)

[52] L Cudmore: *The Centre of Life: A Natural History of the Cell*. (Quadrangle, 1977)

[53] L Wolpert: *The Triumph of the Embryo*. (Oxford University Press, 1991).

Application of feedback control to organisational issues is discussed in [11] and [46-48], its functioning in personal life in

[54] Scott Peck: *The Road Less Travelled*. (Arrow Books, 1990)

and its role in general determination of quality of life in

[55] G F R Ellis: *An Overall Framework for Quality of Life Evaluation, in Social Development in the Third World*, Ed. J G M Hilhorst and M Klatter (Croom Helm, Beckenham, Kent, 1985) 48-62,

[56] G F R Ellis: The dimensions of poverty. *Social Indicators Research* 15: 229-253 (1984).

The dynamic feedback systems controlling the atmosphere and biosphere are described in

[57] J Lovelock: *Gaia*. (Gaia Books Ltd., 1991).

The nature of Ecology is discussed in
[58] E J Kormendy: *Concepts of ecology*. (Prentice-Hall, 1969)
[59] R E Ricklefs: *The Economy of Nature*. (Chiron, 1976).
Practice of agriculture with a long-term future is discussed in
[60] Jim Mollison: *Permaculture One: A Perennial Agriculture for Human Settlements*. (Tagari Publications, 1990)
[61] Jim Mollison: *Permaculture Two: Practical Design for Town and Country*. (Tagari Publications, 1990),
and the nature of long-term resource limitations in
[62] P R Ehrlich and A H Ehrlich: *Population Resources Environment: Issues in Human Ecology*. (Freeman, 1974)
[63] *Managing Planet Earth*, Readings from Scientific American (Freeman, 1990).
The unpredictable behaviour envisaged by chaos theory is discussed in
[64] I Stewart: *Does God Play Dice? The New Mathematics of Chaos*. (Penguin Books, 1990).
The nature of Evolution, and evidence for the theory, are presented in [49] and
[65] R Dawkins: *The Blind Watchmaker*. (Penguin Books, 1986)
[66] M A Edey and D C Johansen: *Blueprints*. (Oxford University Press, 1990)
[67] S M Stanley: *Earth and Life through Time*. (Freeman, 1989)*.
The problem of the origin of life is discussed in [49,65] and
[68] A Scott: *The Creation of Life*. (Blackwell, 1986).

Chapter 5: The Physical Universe

The nature of Astronomical discoveries is presented in
[69] T P Snow: *Essentials of the Dynamic Universe*. (West Publishing Company, St. Paul, 1987)
[70] N Henbest and M Marten: *The New Astronomy*. (Cambridge University Press, 1983).
The historical struggle to understand the solar system is described in
[71] A Koestler: *The Sleepwalkers*. (Penguin, 1959)
and the history of the major discoveries in astronomy in
[72] M Harwit: *Cosmic Discovery: The Search, Scope and Heritage of Astronomy*. (Harvester Press, 1981).
The recent search for understanding of cosmology is presented in
[73] D Overbye: *Lonely Hearts of the Cosmos: The Scientific Quest for the Secret of the Universe*. (Harper Collins, 1991)

[74] A Lightmann and R Brawer: *Origins: The Lives and Worlds of Modern Cosmologists*. (Harvard University Press, 1991).

The nature of the Earth is presented in

[75] J Mitchell (Ed.): *Anatomy of the Earth*. (Mitchell Beazley, 1976).

The forces at action in shaping the Earth's surface are examined in [67] and

[76] V Fuchs (ed.): *Forces of Nature*. (Thames and Hudson, 1977).

Basic astrophysics is discussed in [20] and

[77] I Shlovsky: Stars: *Their Birth Life and Death*. (Freeman, 1978)

[78] F Hoyle and J Narlikar: *The Physics-Astronomy Frontier*. (Freeman, 1980)*.

Cosmology is presented in a masterly survey in

[79] Ted Harrison: *Cosmology: the Science of the Universe*. (Cambridge University Press, 1981),

with recent observations summarised in

[80] J Cornell (Ed.): *Bubbles Voids and Bumps in time: the New Cosmology*. (Cambridge University Press, 1989),

and interesting topics discussed in

[81] J Leslie: *Physical Cosmology and Philosophy*. (MacMillan, 1990).

A general overview of the origins of the solar system and life is given in

[82] J G Eccles: *The Human Mystery*. (Routledge and Kegan Paul, 1984),

[83] A Fabian: *Origins*. (Cambridge University Press, 1988)

(and see also [78,93,95]) while changes in paradigms in cosmology are discussed in

[84] Ted Harrison: *Masks of the Universe*. (Collier Books, New York, 1985)

[85] G F R Ellis: *The Transition to the Expanding Universe. In Modern Cosmology in Retrospect*, Ed. B Bertotti, R Balbinot, S Bergia and A Messina. (Cambridge University Press, 1990), 97.

The physics of the early universe is presented in

[86] D W Sciama: *Modern Cosmology*. (Cambridge University Press, 1971)

[87] S Weinberg: *The First Three Minutes: A Modern View of the Origin of the Universe*. (Basic Books, 1977)

[88] J Silk: *The Big Bang*. (Freeman, 1980).

Recent developments in cosmology are discussed in [20] and

[89] H Pagels: *Perfect Symmetry* (Penguin, 1985)

[90] L Krauss: *The Fifth Essence The Search for Dark Matter in the Universe*. (Vintage, Basic Books, London, 1989)

[91] J Gribbin and M J Rees: *Cosmic Coincidences*. (Black Swan, 1991).

The quantum creation of the universe is discussed in [20] and
[92] S Hawking: *A Brief History of Time*. (Bantam, 1988).
The issue of how many civilisations there are in the universe is considered in
[93] I. S. Shlovsky and C. Sagan: *Intelligent Life in the Universe*. (New York: Dell, 1966).
[94] D Goldsmith: *The Quest for Extraterrestrial Life: a Book of Readings*. (University Science Books, Mill Valley, California, 1980)
[95] J D Barrow and F J Tipler: *The Anthropic Cosmological Principle*. (Oxford, Oxford University Press) 1986*.
The final fate of the universe is discussed in
[96] J N Islam: *The Ultimate Fate of the Universe*. (Cambridge University Press, 1983)
[97] J Gribbin: *The Omega Point*. (Heinemann: 1987).

Chapter 6: Emerging Questions and Uncertainties

Issues of verifiability in cosmology are discussed in
[98] G F R Ellis: *Cosmology and Verifiability. In Physical Sciences and the History of Physics*. Boston Studies in the Philosophy of Science. Volume 82 Ed. R S Cohen and M W Wartofsky. (Reidel, 1984), 93-114.
A proposal for the nature of quantum cosmology is given in [92] (and see also [20]). Olber's paradox is discussed in
[99] E R Harrison: *Darkness at Night: a Riddle of the Universe*. (Harvard University Press, 1987)
(see also [78]) and Mach's Principle in
[100] D W Sciama: *Physical Foundations of General Relativity*. (Heinemann, 1969)
[101] D W Sciama: *The Unity of the Universe*. (Faber, 1959).
The arrow of time is considered in [34] and
[102] P Coveney and R Highfield: *The Arrow of Time*. (Flamingo, 1990).
The uniqueness of the universe is commented on in
[103] W H McCrea: *Rep Prog Phys*. 16 32 (1953)
[104] M K Munitz: *Brit J Philos Sci* 13:104 (1962).

Chapter 7: The Congenial Environment

The Anthropic Principle is discussed in [91], [95] and
[105] P C W Davies: *The Cosmic Blueprint*. (London: Heinemann, 1987)

[106] J Leslie: *Universes*. (Routledge, 1989)
with a masterly survey of the literature of the subject given in
[107] Y V Balashov: Resource Letter Ap-1 *The Anthropic Principle* Amer Journ Phys 54:1069-1076 (1991),
and its relation to statistical testing discussed in
[108] A J M Garrett and P Coles: *Bayesian Inductive Inference and the Anthropic Cosmological Principle*. Preprint (Queen Mary College, London, 1992)*.
The issue of multiples universes is presented in an intriguing novel:
[109] O Stapledon: *Starmaker*. (Dover, 1968).

Chapter 8: The Universe and Humanity

The nature of the human spirit is presented in
[110] W Burnett (Ed.): *The Human Spirit*. (George Allen and Unwin, 1960)
[111] D Cohen: *The Circle of Life*. (Aquarian Press, 1992)
[112] J Campbell with Bill Moyers: *The Power of Myth*. (Doubleday, 1988)
while, additional to the great literature from the Greek tragedies through Shakespeare to Dostoyevsky and beyond, three documents movingly portraying humanity are
[113] T S Eliot: *Four Quartets*. (Faber, 1970)
[114] A St.-Exupery: *Flight to Arras*. (Heinemann, 1955)
[115] J Agee: *Let Us Now Praise Famous Men*. (Houghton-Mifflin, 1960).
Practical Psychology is discussed in [54] and
[116] Victor Frankl: *Man's Search for Meaning*. (Hodder and Stoughton, 1991)
[117] Erich Fromm: *To Have and To Be*. (Abacus, 1987)
[118] A Storr: *The Integrity of the Personality*. (Penguin, 1960).
The nature of Society is presented in
[119] D F Aberle, A K Cohen, A K Davis, M J Levy and F X Sutton: *The Functional Pre-requisites of a Society*. Ethics, Vol 60, (1950); Bobbs-Merrill Reprints in the Social Sciences S-1,
[120] Peter Berger: *Invitation to Sociology*. (Doubleday, 1963),
and the way we perceive social reality in
[121] Peter Berger and T Luckmann: *The Social Construction of Reality*. (Penguin, 1979).

Intimations of transcendence and the significance of the individual are discussed in

[122] Peter Berger: *A Rumour of Angels*. (Doubleday, 1970)
[123] W Stoeger: *Cosmology and Theology: the Avenues of their Mutual Interaction*. To appear, Torino conference proceedings (Ed. A. Curir) (1992)
[124] C S Lewis: *Mere Christianity*. (Fontana, 1975).

The rationalist view is proposed inter alia in [65] and

[125] J Monod: *Chance and Necessity*. (Fontana, 1972).

Different theories of the nature of mankind are presented clearly in

[126] L Stevenson: *Seven Theories of Human Nature*. (Oxford University Press, 1974)

and the socio-biology perspective in

[127] T H Clutton-Brock and P H Harvey: *Readings in Socio-biology*. (Freeman, 1978).

The nature of Religious experience is surveyed in

[128] William James: *The Varieties of Religious Experience*. (Fontana, 1960)
[129] F C Happold: *Mysticism*. (Penguin, 1965)
[130] Jean Hardy: *Psychology with a Soul*. (Arkana, 1987),

and the nature of the Christian religious approach in

[131] H Kung: *Does God Exist?* (Collins, 1979)
[132] Stephen Neill: *A Genuinely Human Existence*. (Constable, 1959)
[133] R. N. Bellah: *Beyond Belief*. (Harper and Row, 1976).

Its basis in particular experiences is illustrated by

[134] A Frossard: *God Exists for I Have Met Him*. (Collins, 1970)
[135] Lin Yutang: *From Pagan to Christian*. (Heinemann, 1960)
[136] C S Lewis: *Surprised by Joy*. (Geoffrey Bles, 1955).

The Quaker view is presented in

[137] G Hubbard: *Quaker by Convincement*. (Penguin, 1985)
[138] *Christian Faith and Practice in the Experience of the Society of Friends*. (London Yearly Meeting of the Religious Society of Friends, Friends House, Euston Road, London, 1972).

The fundamental nature of the sacrificial view of ethical behaviour is portrayed in

[139] W Temple: *Readings in St. John's Gospel*. (St Martins Library, 1961)

and its working out in practice in

[140] Martin Luther King: *Strength to Love*. (Collins, 1986)

[141] R Duncan: *The Writings of Ghandi.* (Fontana, 1983).

A natural theology of cosmology is given in

[142] P C W Davies: *God and the New Physics.* (Pelican, 1984)

[143] P C W Davies: *The Mind of God.* (New York: Simon and Schuster, 1992),

while a specifically Christian view is given in

[144] G F R Ellis: *The Theology of the Anthropic Principle* (1991). To appear in [6],

and is discussed further in

[145] N Murphy: *Evidence of Design in the Fine-tuning of the Universe* (1992). To appear in [6].

An overview of human nature and intelligence is given in

[146] J Eccles and D N Robinson: *The Wonder of Being Human: Our Brain and Mind.* (New Science Library, 1985)

and the basis of consciousness discussed in

[147] E Harth: *Windows on the Mind: Reflections on the Physical Basis of Consciousness.* (Harvester Press, 1982)

[148] C Blakemore and S Greenfield (Ed.) *Mindwaves.* (Blackwell, 1987)

[149] G M Edelman: *Bright Air, Brilliant Fire: On the Matter of Mind.* (Basic Books, 1992)

Chapter 9: Conclusion

The ecological Implications of a responsible view are presented in [62,63], while the kind of life implied by the religious world view is portrayed in [140,141] and

[150] Alan Paton: *Instrument of Thy Peace.* (Seabury, 1968)

(and see also [124,132,138]).

NOTES

Chapter 1.
1. The abbreviation ' i.e.' = 'that is' will be used to indicate a brief definition of a term that may be unfamiliar to some of my readers, or to establish the particular connotation I have in mind for a word.

Chapter 2.
1. I assume this reality exists, and can to some extent be discovered. There are of course various viewpoints that deny one or other of these assumptions. While they may be philosophically amusing, they cannot be sustained in a serious search for a satisfying overall view of the nature of the world, precisely because they deny that that quest can have any meaning.
2. An alternative viewpoint is that physical behaviour is not governed by such laws, but accurately described by them. From the practical standpoint, the difference between these views is immaterial.
3. This does not necessarily mean we can describe the ultimate *nature* of that reality, but rather that we can characterise its *effective nature*, which as far as we are concerned, is absolute; for example, if people have no food, they will die.
4. The exception here is recent use of computers to determine the nature of mathematical systems, and in particular to examine the behaviour of chaotic and fractal systems, where some of mathematics has begun to take on aspects of an experimental science.
5. Assuming the fundamental laws were the same then as now. This basic assumption is to a certain extent susceptible to test.
6. There is a trivial relativisation in terms of the language used to describe reality, for example "oxygen" and *sauerstoff* are the same thing; I assume we are able to handle such issues of multiple representation without getting misled by them.
7. The impression that relativity theory has done away with such a reality

is incorrect: that theory emphasizes both the relativity of different observer's views of that reality and the possibility of characterising its nature in an invariant way [31].

8. See [10] for a discussion of the case of cosmology.

9. I return to this theme in Chapter 8.

Chapter 3.

1. See *Powers of Ten* [17] for a superb overview of the relative sizes of things.

2. This is plural; the singular form is *a quantum of energy.*

Chapter 4.

1. It is convenient to use this language for living systems, as well as for man-made ones, without prejudging the issue of whether there is a Designer or not; this issue will be discussed later. It clearly makes sense to say that a wing is designed so that a bird can fly.

2. On the other hand, quantum mechanics does have this implication, and could be the source of unpredictable fluctuations that are amplified to macroscopic scales in a chaotic system.

Chapter 5.

1. In the following, I emphasize the unique identity of our own Earth, Sun, and Galaxy by using capital letters in their names.

2. We here refer to temperatures in degrees Kelvin, denoted by 'K'; this denotes degrees above absolute zero (which is - 273 degrees Centigrade). Thus 0 K is absolute zero, the lowest temperature that can exist.

3. The relative sizes and distances of the objects in our environment are illustrated dramatically in *Powers of Ten* [17].

4. About 54 thousand billion billion km., compared with the 150 million km. from the Earth to the Sun.

5. The alternative possibility is that we are at the centre of the Universe, which is spherically symmetric about us. This is consistent with the observations, but (for philosophical reasons) is believed to be unlikely [81].

6. Gravitational redshifts can occur in a static universe, but observations will appear isotropic about us only if we are at the centre of the universe; this is possible, but is not philosophically popular as an idea.

7. In principle a *cosmological constant,* that is a uniform force of repulsion proportional to distance and independent of the nature of matter,

could avoid this conclusion, but in practice evidence related to the observed background radiation and the largest observed redshifts shows this remedy will not work.

8. The heavy forms of hydrogen.

9. Indeed they might have decided to save themselves the time and expense of doing so, precisely because one of their logicians had come up with this very argument, and thereby proved there was no purpose in exploring the Galaxy!

10. It should be called a *cold death,* but this name became established in the 1930's when the possibility was first discovered.

11. This is the upper case (Capital) Greek letter Omega.

12. Recently two groups have claimed astronomical evidence for high densities, but both claims have been disputed, and neither is generally accepted. Very recently new evidence (based on analysing galaxy motions) has come in suggesting the high density hypothesis is correct; this has still to be evaluated by other workers in the field.

Chapter 6.

1. It is understood that "light" is a generic term for any form of electro-magnetic radiation by which we can see distant objects: radio waves, infra-red radiation, ultraviolet radiation, and X-rays as well as ordinary light [28].

2. See the discussion of Relativity in Chapter 3.

3. Which later will become galaxies.

4. There are also particle (causal) horizons if the inflationary Universe has infinite spatial sections; but they do not exist if it has closed spatial sections (then the particle horizons are broken at very early times, when light has had time to travel right round the Universe).

5. Apart from what happens to matter in the final state of collapse in a black hole; but that is completely inaccessible to observation.

6. While this has originally been developed as a quantum cosmology idea, it is now known there are classical solutions of the Einstein equations with the same property.

7. Updated versions of the calculation determine what background radiation we expect to receive at each wavelength, from vary distant matter in the Universe; this can then be compared with observations.

8. Einstein developed his static Universe model (1917) in the hope that it would show General Relativity fully incorporates Mach's principle in its structure.

However, de Sitter's universe model (also found in 1917) showed this is not so.

9. There is an exception in the case of some of the weak interactions; this is difficult to detect, and it seems very unlikely it is the fundamental source of the arrow of time.

10. Quite apart from the issue of chaotic dynamics, briefly mentioned in Chapter 4.

11. Cf. the discussion in Chapter 2.

12. As has been emphasized elsewhere in this book, we can attain effective certainty in numerous practical matters, controlled by the regular action and effectiveness of physical laws.

Chapter 7.

1. Thus local systems do depend critically on the large-scale properties of very distant matter; but not on its detailed properties, so allowing locality. The large-scale structure provides a stable local environment within which forces can act to provide locally deterministic behaviour.

2. I omit the so-called Final Anthropic Principle (FAP for short [95]), which maintains that intelligent life must necessarily evolve and then remain in existence until the end of the universe, for I do not believe it merits serious discussion as a scientific proposal; indeed it led to a famous book review referring to the Completely Ridiculous Anthropic Principle (CRAP for short).

3. Discussed below.

4. This is one of the few real alternatives proposed to the Copenhagen interpretation of quantum mechanics, which leads to the necessity of an Observer, and so to the controversial Strong Anthropic interpretation considered above.

5. There will be some who will reject this possibility out of hand, as meaningless or as unworthy of consideration. However it is certainly logically possible. Simple rejection without considering the arguments for and against is evidence of a closed mind, unable to contemplate the complete range of possibilities (cf. the discussion in Chapter 2).

6. Life Force, God, or whatever other name might be attributed to this underlying facet of reality. In what follows it will often be convenient to use the word "God", without committing the discussion to a particular viewpoint on the specific meaning of that concept.

7. One can contemplate properties of families of physical laws, but one can only organise this investigation by proposing higher level laws (laws for laws of physics!); immediately the issue then is, so what controls these higher level laws?

Chapter 8.

1. As happens also in the "hard" sciences at the limits of its understanding at any particular time; but in that case, many of these uncertainties are resolved as science progresses.

2. This was not always so, but is now effectively universal; in my view this should be taken as an indication of moral evolution, as humanity has, over time, come to terms with consciousness, and become more adept at perceiving the nature of underlying moral order.

3. We cannot scientifically prove that anything is valuable; all we can do is prove something contributes to some other purpose we regard as valuable, which only helps us if we can prove that this other purpose in turn is valuable; this is precisely what science cannot do.

4. In the latter case, we just assert that these are the correct values without any further justification.

5. As are the foundations of logic and of mathematics (the logical sciences that are not susceptible to testing or proof).

6. The charge that theology might be a theory without an object is not particularly worrying. It applies equally to major tracts of modern theoretical physics: string theory, supersymmetry, cosmic strings, quantum cosmology to name a few.

7. But not necessarily with the detailed understandings of science, which may keep changing and is therefore an unfirm foundation.

8. It is my belief the Christian religion [122-124,131-136] (and in that tradition, the Quaker view [137,138]) has shown more of that truth than any other, but that does not deny that the others too have valuable insights and visions of underlying reality. See [135] for an enlightening contrast of Eastern religions and Christianity.

9. With just one exception: this sentence itself.

10. This is part of *the problem of evil*: a full Cosmology must have some reasonable account of evil as well as of good. (This will be briefly considered later.)

11. While other theological evidence [5], such as tradition and scriptures, is essential in terms of providing our only access to historical evidence and past experience, in fundamental terms of access to underlying reality it is secondary to this primary source, so will not be considered here.

12. I hesitate to use the word "spiritual", because of the abuse it has suffered, but no other word will convey what is intended here.

13. This view goes considerably further than the natural theology of [142,143], which concentrates only on physical understanding.

14. Here we do so in fairly general terms. A more specifically Christian analysis is given in [144,145].

15. A physical basis for this is possible even within the context of known physics because of the lack of determinism in microphysical behaviour resulting from the quantum uncertainty principle (which might also be related to the existence of free will [6]).

16. On the Quaker view, referred to as an "inner light" in each one of us [137,138].

INDEX